Enhanced Virtual Prototyping

Vladimir Herdt • Daniel Große • Rolf Drechsler

Enhanced Virtual Prototyping

Featuring RISC-V Case Studies

Vladimir Herdt
University of Bremen and DFKI GmbH
Bremen, Germany

Daniel Große
Johannes Kepler University Linz
Linz, Austria

Rolf Drechsler
University of Bremen and DFKI GmbH
Bremen, Germany

ISBN 978-3-030-54830-8 ISBN 978-3-030-54828-5 (eBook)
https://doi.org/10.1007/978-3-030-54828-5

This Springer imprint is published by the registered company Springer Nature Switzerland AG
The registered company address is: Gewerbestrasse 11, 6330 Cham, Switzerland

To Elena and Oleg,
Lukas
and
Lothar

Preface

Virtual Prototypes (VPs) play a very important role to cope with the rising complexity in the design flow of embedded devices. A VP is essentially an executable abstract model of the entire *Hardware* (HW) platform and pre-dominantly created in SystemC TLM (*Transaction Level Modeling*). In contrast to a traditional design flow, which first builds the HW and then the *Software* (SW), a VP-based design flow enables parallel development of HW and SW by leveraging the VP for early SW development and as reference model for the subsequent design flow steps. However, this modern VP-based design flow still has weaknesses, in particular due to the significant manual effort involved in verification and analysis as well as modeling tasks which is both time-consuming and error-prone. This book presents several novel approaches that cover modeling, verification, and analysis aspects to strongly enhance the VP-based design flow. In addition, the book features several RISC-V case studies to demonstrate the developed approaches. The book contributions are essentially divided into four areas: The first contribution is an open-source RISC-V VP that is implemented in SystemC TLM and covers functional as well as non-functional aspects. The second contribution improves the verification flow for VPs by considering novel formal verification methods and advanced automated coverage-guided testing techniques tailored for SystemC-based VPs. The third contribution are efficient coverage-guided approaches that improve the VP-based SW verification and analysis. The fourth and final contribution of this book are approaches that perform a correspondence analysis between *Register-Transfer Level* (RTL) and TLM to utilize the information available at different levels of abstraction. All the approaches are presented in detail and are extensively evaluated with several experiments that clearly demonstrate their effectiveness in strongly enhancing the VP-based design flow.

Bremen, Germany — Vladimir Herdt
Bremen, Germany — Daniel Große
Bremen, Germany — Rolf Drechsler

Acknowledgement

First, we would like to thank the members of the research group for *Computer Architecture* (AGRA) at the University of Bremen as well as the members of the research department for *Cyber-Physical Systems* (CPS) of the *German Research Center for Artificial Intelligence* (DFKI) at Bremen. We appreciate the great atmosphere and stimulating environment. Furthermore, we would like to thank all the co-authors of the papers which formed the starting point for this book: Hoang M. Le, Muhammad Hassan, Mingsong Chen, Jonas Wloka, Tim Güneysu, and Pascal Pieper. We especially thank Hoang M. Le and Muhammad Hassan for many interesting discussions and successful collaborations. Last but certainly not least, special thanks go to Rudolf Herdt for numerous inspiring and helpful discussions.

Bremen, Germany Vladimir Herdt

Bremen, Germany Daniel Große

Bremen, Germany Rolf Drechsler

May 2020

Contents

List of Algorithms

List of Figures

List of Tables

Chapter 1
Introduction

Embedded systems are prevalent nowadays in many different application areas ranging from *Internet-of-Things* (IoT) to automotive and production as well as communication and multi-media applications. Embedded systems consist of *Hardware* (HW) and *Software* (SW) components and are typically small resource constrained systems that are highly specialized to implement application specific solutions. Hence, design flows for embedded systems require efficient and flexible design space exploration techniques to satisfy all application specific requirements such as power consumption and performance constraints.

In addition, the complexity of embedded systems is continuously increasing. They integrate more and more complex IP (*Intellectual Property*) components (such as processor cores, custom accelerators, and other HW peripherals) on the HW side and extensively rely on SW to control and access the IPs as well as to perform other important functionality. In particular the SW complexity has been rising very strongly in the recent years and thus SW should be considered with equal priority to the HW in the design process. Nowadays, the (embedded) SW encompasses multiple abstraction layers ranging from bootcode to device drivers up to full blown operating systems in combination with several libraries (such as a wireless network stack for communication) in addition to the actual application code. Furthermore, due to its flexibility, SW is also increasingly leveraged to consider non-functional aspects as well, such as implementing strategies for controlling the power management of the embedded system. Besides developing the SW, verification of the SW is also crucial to avoid errors and security vulnerabilities. Similar to the SW development time, the SW verification time increases as well with the rising SW complexity. Therefore, it is very important to start as early as possible with the SW development and verification to meet the tight time-to-market constraints and deliver a high-quality product.

The traditional design flow for embedded systems is insufficient to cope with this rising complexity. The traditional design flow works in a sequential fashion by first developing the HW and then the SW (as shown on the left side of Fig. 1.1). The

V. Herdt et al., *Enhanced Virtual Prototyping*,
https://doi.org/10.1007/978-3-030-54828-5_1

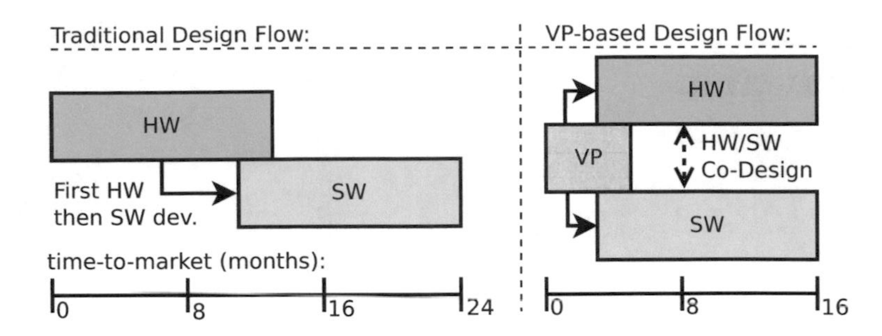

Fig. 1.1 Comparison of the concept of a traditional design flow (left side) and VP-based design flow (right side)

reason is that the SW requires the HW to be executed, hence SW development starts once the HW is almost complete. This in turn results in significant delays in the design flow in particular due to the rising SW complexity.

To deal with this issue, *Virtual Prototypes* (VPs) are increasingly leveraged for SW execution early in the design flow and thus enable parallel development of HW and SW [26, 31, 46, 139, 189]. A VP is essentially an executable abstract model of the entire HW platform and pre-dominantly created in SystemC TLM (*Transaction Level Modeling*) [113, 163].[1] From the SW perspective a VP mimics the real HW, i.e. it describes the HW at a level that is relevant for the SW. For example, a VP provides all SW visible HW registers but abstracts away from complex internal communication protocols. However, being an abstract model of the real HW, the VP can be developed much faster than the HW. Once the VP has been created, it can be used to execute the SW and hence enables SW development and testing early in the design flow. At the same time the VP also serves as executable reference model for the subsequent HW design flow. Thus, a VP-based design flow enables parallel development of HW and SW (as shown on the right side of Fig. 1.1), which leads to a significant reduction in time-to-market and also enables a more agile development flow by passing feedback between the HW and SW development flows.

In the following, first the VP-based design flow is described in more detail and then the contributions of this book that improve the VP-based design flow are presented.

[1]Essentially, SystemC is a C++ class library that includes an event-driven simulation kernel and provides common building blocks to facilitate the development of VPs, while TLM enables the description of communication in terms of abstract transactions. More details on SystemC TLM will be provided in later in the preliminaries Sect. 2.1.

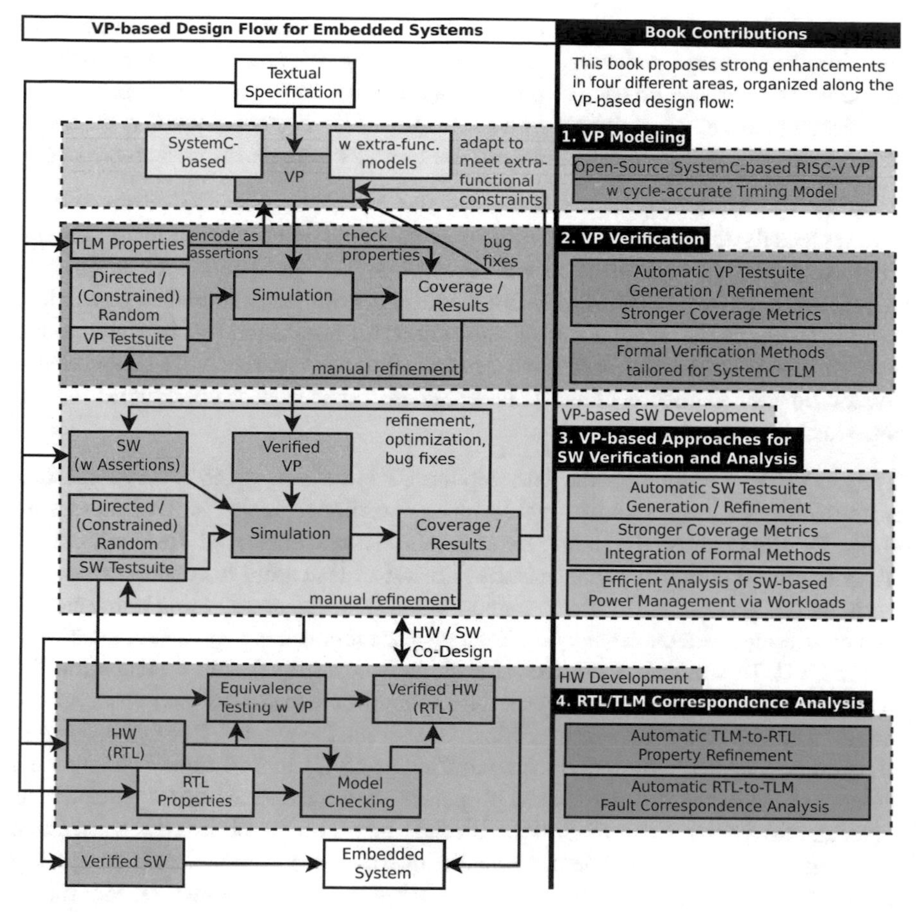

Fig. 1.2 Overview on the VP-based design flow (left side) and the corresponding contributions of this book (right side)

1.1 Virtual Prototype-Based Design Flow

Figure 1.2 (on the left side) shows the VP-based design flow in more detail. Starting point is a textual specification of the embedded system that specifies the functional as well as non-functional requirements for the embedded system. The VP-based design flow itself is essentially separated into four different steps (highlighted with a different background color):

1. VP Modeling (yellow)
2. VP Verification (green)
3. VP-based SW Development (blue)
4. HW Development (red)

Please note, these four design flow steps are not performed sequentially one after another but are interleaved, as, for example, the SW and HW are developed in parallel. An overview on each step (one after another from top to bottom, grouped by dashed boxes and highlighted with the respective background color) is shown on the left side of Fig. 1.2. In the following, the four steps are described in more detail.

Step 1: VP Modeling In the first step, the SystemC-based VP is created. A VP represents the entire HW platform. It can consist of one or many general or special purpose processors as well as several HW peripherals. In addition to functional behavior, typically non-functional behavior models are integrated into the VP to obtain early and accurate estimations on non-functional behavior of the system alongside the SW execution. These estimations enable early design space exploration and validation of non-functional constraints such as power consumption and execution performance/timing.

Step 2: VP Verification In the next step the VP is verified which is very important because the VP serves as executable reference model in the subsequent design flow. Pre-dominantly simulation-based methods are employed for verification. They require a comprehensive testsuite (i.e. set of testcases) to ensure a thorough verification. Testcases are either created manually or generated using pure random and constrained random techniques. Each testcase represents a specific input stimuli for the VP. To simulate a testcase, a testbench is set up that instantiates the VP configuration under test, then passes the input stimuli to the VP and observes the output behavior of the VP. It is possible to instantiate the whole VP or test individual VP components (e.g. the interrupt controller or CPU) in isolation. The expected behavior is specified in form of TLM properties which are either directly encoded as (executable) assertions into the VP or checked based on the observed results. Code coverage information (i.e. line and branch coverage) are used to access the quality of the testsuite and guide the testcase generation process. However, significant manual effort is required to create directed testcases that maximize the coverage. The verified VP serves as common executable reference model for the subsequent SW and HW development and enables parallel development of the SW and HW.[2]

Step 3: VP-based SW Development The primary use case for the VP is to enable *early* SW development. Besides SW development and design space exploration, the VP is also used to perform additional extensive SW verification and analysis tasks. Similar to the VP verification itself, again pre-dominantly simulation-based methods are employed. However, these simulation-based methods operate on a different abstraction level, since they are designed to primarily test the SW and not the VP. Hence each testcase represents input stimuli for the SW. SW properties

[2]However, please note, that due to the complexity of the VP verification task and significant manual effort involved, only a preliminary VP verification is performed in the beginning and the VP verification task is typically an ongoing process alongside the SW and HW development. Therefore, improvements in the VP verification area are very important to avoid propagation of errors and thus costly iterations.

are either encoded as assertions into the SW itself or checked based on the observed output behavior of the VP. The testcase generation process is again mostly based on manually created directed testcases as well as (constrained) random techniques and guided by the observed results and code coverage information (though in this case the SW coverage is considered). Similar to verifying the VP, significant time consuming and error prone manual effort are required to perform an extensive verification/analysis of the SW.

Based on the obtained results the SW is refined and optimized to fix bugs and meet the application specific performance and power consumption constraints. In particular in the early phases of the design flow it is also possible to re-configure the VP in order to optimize the HW to meet the application specific requirements.

Step 4: HW Development The HW development takes place in parallel to the SW development. The result is a synthesizable *Register-Transfer Level* (RTL) description of the HW. Mostly two different verification methods are employed:

- Perform an equivalence testing of the HW and the VP by re-using the SW testsuite(s) obtained in step 3.
- Perform model checking based on RTL properties that are created based on the textual specification.

Finally, the verified SW and HW can be integrated to the final embedded system. Please note, HW development and HW verification methods at RTL and below are not the focus of this book.

1.2 Book Contribution

The VP-based design flow significantly improves on the traditional design flow and has also been proven successful on several industrial use cases [46, 124, 126, 160]. However, this modern VP-based design flow still has weaknesses, in particular due to the extensive manual effort involved for verification and analysis tasks which is both time consuming and error prone. This book proposes several novel approaches to strongly enhance the VP-based design flow and with this provides contributions to every of the four main steps in the VP-based design flow. An overview of the book contributions is shown on the right side of Fig. 1.2. The book contributions are grouped in four areas which (mostly) correspond to the four steps of the VP-based design flow and will be detailed in the following:

1. VP Modeling
2. VP Verification
3. VP-based Approaches for SW Verification and Analysis
4. RTL/TLM Correspondence Analysis

Contribution Area 1: VP Modeling The first contribution of this book is an open-source RISC-V VP implemented in SystemC TLM.

RISC-V is an open and free *Instruction Set Architecture* (ISA) [218, 219] which is license-free and royalty-free. Similar to the enormous momentum of open-source SW the open-source RISC-V ISA is also gaining large momentum in both industry and academia. In particular for embedded devices, e.g. in the IoT area, RISC-V is becoming a game changer. A large and continuously growing ecosystem is available around RISC-V ranging from several HW implementations (i.e. RISC-V cores) to SW libraries, operating systems, compilers, and language implementations. In addition, several open-source high-speed *Instruction Set Simulators* (ISS) are available. However, these ISSs are primarily designed for a high simulation performance and hence can hardly be extended to support further system-level use cases such as design space exploration, power/timing/performance validation, or analysis of complex HW/SW interactions. The goal of the proposed RISC-V VP is to fill this gap in the RISC-V ecosystem and stimulate further research and development.

The VP provides a 32/64 bit RISC-V core with an essential set of peripherals and support for multi-core simulations. In addition, the VP also provides SW debug (through the Eclipse IDE) and coverage measurement capabilities and supports the FreeRTOS, Zephyr, and Linux operating system. The VP is designed as extensible and configurable platform (as an example we provide a configuration matching the RISC-V HiFive1 board from SiFive) with a generic bus system and implemented in standard-compliant SystemC. The latter point is very important, since it allows to leverage cutting-edge SystemC-based modeling techniques needed for the mentioned system-level use cases. Finally, the VP allows a significantly faster simulation compared to RTL, while being more accurate than existing ISSs.

In addition, the VP integrates an efficient core timing model to enable fast and accurate performance evaluation for RISC-V based systems. The timing model is attached to the core using a set of well-defined interfaces that decouple the functional from the non-functional aspects and enable easy re-configuration of the timing model. As example a timing configuration matching the RISC-V HiFive1 board from SiFive is provided.

Contribution Area 2: VP Verification This book improves the VP verification flow by considering novel formal verification methods and advanced automated coverage-guided testing techniques tailored for SystemC-based designs.

Formal verification methods can prove the correctness of a SystemC design with respect to a set of properties. However, formal verification of SystemC designs is very challenging as it has to consider all possible inputs as well as process scheduling orders which can make the verification process intractable very quickly due to the state space explosion problem. In this book advanced symbolic simulation techniques for SystemC have been developed to mitigate state space explosion. Besides the basic symbolic simulation technique that enables the foundation for efficient state space exploration, this book presents several major extensions and optimizations, such as SSR (*State Subsumption Reduction*) and CSS (*Compiled Symbolic Simulation*). SSR provides support for verification of cyclic state spaces by preventing revisiting symbolic states and therefore making the verification complete. CSS is a complementary technique that tightly integrates the symbolic simulation engine with the SystemC design under verification to drastically boost the verification performance by enabling native execution support. While the developed formal techniques have significantly improved the existing state-of-the-art on formal verification of SystemC designs, they nonetheless are still susceptible to state space explosion. Formal methods are applicable to single VP components in isolation, for example, to verify an interrupt controller, as this book will show later on, but are still not applicable to verify the whole VP.

Therefore, this book also considers advanced coverage-guided testing techniques that rely on testcase generation and simulation. Compared to the existing simulation-based verification flow this book investigates stronger coverage metrics as well as advanced automated testcase generation and refinement techniques. In particular it considers the *Data Flow Testing* (DFT) coverage and *Coverage-guided Fuzzing* (CGF), both of which have been very effective in the SW domain, tailored for verification of VPs. For DFT, a set of SystemC specific coverage criteria have been developed that consider the SystemC semantics of using non-preemptive thread scheduling with shared memory communication and event-based synchronization. CGF is applied for verification of *Instruction Set Simulators* (ISSs), i.e. an abstract model of a processor core, and is further improved by integrating functional coverage and a custom mutation procedure tailored for ISS verification.

Contribution Area 3: VP-based Approaches for SW Verification and Analysis
First, this book proposes novel VP-based approaches for SW verification. They improve the existing VP-based SW verification flow by integrating stronger coverage metrics and providing automated testcase generation techniques as well as leverage formal methods. Ensuring correct functional behavior is very important to avoid errors and security vulnerabilities (such as buffer overflows).

The first proposed approach integrates concolic testing into the VP. Essentially, concolic testing is an automated technique that successively explores new paths through the SW program by solving symbolic constraints that are tracked alongside the concrete execution. However, an integration of concolic testing with VPs is challenging due to complex HW/SW interactions and peripherals. The proposed solution enables a high simulation performance with accurate results and comparatively little effort to integrate peripherals with concolic execution capabilities.

While very effective, concolic testing relies on symbolic constraints and thus is susceptible to scalability problems. Therefore, this book proposes a second SW verification approach that leverages state-of-the-art CGF in combination with VPs for verification of embedded SW binaries. To guide the fuzzing process the coverage from the embedded SW is combined with the coverage of the SystemC-based peripherals of the VP. This second approach does not rely on formal methods, and hence does not suffer from extensive scalability problems, and at the same time yields much better results than (constrained) random testing due to the involved automated coverage feedback loop.

Beside correct functional behavior, a high performance in combination with low power consumption is a key requirement for many embedded systems. Due to its ease of use and flexibility power management strategies are often implemented in

SW. PM strategies are analyzed by observing the non-functional behavior of the VP alongside the SW execution.

This book proposes novel VP-based techniques tailored for validation of SW-based power management strategies. The contribution is two-fold: First, a constrained random encoding to specify abstract workloads that enables generation of a large testsuite with different application workload characteristics. And second, an automated coverage-guided approach to generate workloads that maximize the coverage with respect to the power management strategies.

Contribution Area 4: RTL/TLM Correspondence Analysis The final contribution of this book is two approaches that perform a correspondence analysis between TLM and RTL.

The first proposed approach enables an automated TLM-to-RTL property refinement. It enables to transform high-level TLM properties into RTL properties to serve as starting point for RTL property checking. This avoids the manual transformation of properties from TLM to RTL which is error prone and time consuming.

The second proposed approach performs an RTL-to-TLM fault correspondence analysis. It enables to identify corresponding TLM errors for transient bit flips at RTL. The obtained results can improve the accuracy of a VP-based error effect simulation. An error effect simulation essentially works by injecting errors into the VP during SW execution to check the robustness of the SW against different HW faults. Such an analysis is very important for embedded systems that operate in vulnerable environments or perform safety critical tasks to protect against effects of, for example, radiation and aging.

Wrap-Up All the aforementioned contributions have been implemented and extensively evaluated with several experiments. Detailed description and results will be presented in the following chapters. In summary, these contributions strongly enhance the VP-based design flow as demonstrated by the experiments. One of the main benefits is the drastically improved verification quality in combination with a significantly lower verification effort due to the extensive automatization. On the one hand, this reduces the number of undetected bugs and increases the overall quality of the embedded system. On the other hand, this shortens the time-to-market.

1.3 Book Organization

First, Chap. 2 provides relevant background information on SystemC TLM, RISC-V as well as CGF and symbolic execution. Then, Chap. 3 presents the open-source RISC-V VP as well as the integrated timing model that enables a fast and accurate performance evaluation for RISC-V. Next, Chap. 4 presents formal verification techniques using symbolic simulation for SystemC-based VPs. Chapter 5 presents complementary coverage-guided testing techniques for VPs to avoid the scalability issues of formal verification. The next two chapters describe VP-based approaches. Chapter 6 presents VP-based approaches that employ concolic testing and CGF for SW verification. Then, Chap. 7 presents a coverage-guided VP-based approach for validation of SW-based power management. The approach employs constraints and a refinement procedure tailored for generation of specific SW application workloads. Chapter 8 presents two approaches that perform a correspondence analysis between RTL and TLM. In particular, an automated TLM-to-RTL property refinement and an RTL-to-TLM fault correspondence analysis. Finally, Chap. 9 concludes the book.

Chapter 2
Preliminaries

To keep the book self-contained and avoid duplication this section provides background information on relevant and common topics. First, SystemC TLM is introduced in Sect. 2.1. SystemC TLM is used throughout the whole book since it is the language of choice to create VPs. Then, the main concepts of the RISC-V ISA are described in Sect. 2.2. RISC-V is used in several evaluations and case studies in this book and is the ISA implemented in our proposed open-source RISC-V based VP. Finally, Sects. 2.3 and 2.4 briefly introduce the main concepts of CGF and symbolic execution. Both are very effective techniques for SW testing and verification and serve as foundation for several verification approaches developed in this book.

2.1 SystemC TLM

SystemC is a C++ class library that includes an event-driven simulation kernel and provides common building blocks to facilitate the development of VPs for embedded systems.

The structure of a SystemC design is described with ports and modules, whereas the behavior is described in processes which are triggered by events and communicate through channels and in particular TLM transactions. SystemC provides three types of processes with SC_THREAD being the most general type, i.e. the other two can be modeled by using SC_THREAD. A process gains the runnable status when one or more events of its sensitivity list have been notified. The simulation kernel selects one of the runnable processes and executes this process non-preemptively. The kernel receives the control back if the process has finished its execution or blocks itself by executing a context switch. A context switch is either one of the function calls *wait(event)*, *wait(time)*, *suspend(process)*. They will be

V. Herdt et al., *Enhanced Virtual Prototyping*,
https://doi.org/10.1007/978-3-030-54828-5_2

briefly discussed in the following. Basically SystemC offers the following variants of *wait* and *notify* for event-based synchronization:

- *wait(event)* blocks the current process until the notification of the event.
- *notify(event)* performs an *immediate notification* of the event. Processes waiting on this event become immediately runnable in this *delta cycle*.
- *notify(event, delay)* performs a *timed notification* of the event. It is called a *delta notification* if the delay is zero. In this case the notification will be performed in the next *delta phase*, thus a process waiting for the event becomes runnable in the next *delta cycle*.
- *wait(delay)* blocks the current process for the specified amount of time units. This operation can be equivalently rewritten as the following block { `sc_event e; notify(e, delay); wait(e);` }, where *e* is a unique event.

More informations on *immediate*-, *delta*- and *timed*-notifications will be presented in the next section which covers the simulation semantics of SystemC.

Additionally, the *suspend(process)* and *resume(process)* functions can be used for synchronization. The former immediately marks a process as suspended. A suspended process is not runnable. The *resume* function unmarks the process again. It is a form of delta notification, thus its effects are postponed until the next *delta* phase of the simulation. Suspend and resume are complementary to event-based synchronization. Thus a process can be suspended and waiting for an event at the same time. In order to become runnable again, the process has to be resumed again and the event has to be notified.

2.1.1 TLM-Based Communication

Communication between SystemC modules is primarily implemented through TLM transactions. A transaction object essentially consists of the command (e.g. read/write), the start address, the data length, and the data pointer. It allows implementing various memory access operations. Optionally, a transaction can be associated with a delay (modeled as *sc_time* data structure), which denotes the execution time of the transaction and allows to obtain a more accurate overall simulation time estimation.

Figure 3.8 shows a basic sensor model implementation in SystemC that communicates through TLM transactions (the *transport* method) to demonstrate the SystemC TLM modeling principles. We will describe the example in more detail later in Sect. 3.1.5.1.

2.1.2 Simulation Semantics

The execution of a SystemC program consists of two main steps: an *elaboration* phase is followed by a *simulation* phase. During *elaboration* modules are instantiated, ports, channels and TLM sockets are bound and processes registered to the simulation kernel. Basically elaboration prepares the following simulation. It ends by a call to the *sc_start* function. An optional maximum simulation time can be specified. The simulation kernel of SystemC takes over and executes the registered processes. Basically simulation consists of five different phases which are executed one after another [113].

1. *Initialization*: First the *update* phase as defined in step 3 is executed, but without proceeding to the subsequent *delta notify* phase. Then all registered processes, which have not been marked otherwise, will be made runnable. Finally the *delta notify* phase as defined in step 4 is carried out. In this case it will always proceed to the *evaluation* phase.
2. *Evaluation*: This phase can be considered the main phase of the simulation. While the set of runnable processes is not empty an arbitrary process will be selected and executed or resumed (in case the process had been interrupted). The order in which processes are executed is arbitrary but deterministic.[1] Since process execution is not preemptive, a process will continue until it terminates, executes a wait statement or suspends itself. In either case the executed process will not be runnable. Immediate notifications can be issued during process execution to make other waiting process runnable in this evaluation phase. Once no more process is runnable, simulation proceeds to the *update* phase.
3. *Update*: Updates of channels are performed and removed. These updates have been requested during the evaluation phase or the elaboration phase if the *update* phase is executed (once) as part of the initialization phase. The evaluation phase together with the update phase corresponds to a *delta cycle* of the simulation.
4. *Delta Notify*: Delta notifications are performed and removed. These have been issued in either one of the preceding phases. Processes sensitive on the notification are made runnable. If at least one runnable process exists at the end of this phase, or this phase has been called (once) from the *initialization* phase, simulation continues with step 2.
5. *Timed Notification*: If there are timed notifications, the simulation time is advanced to the earliest one of them. If the simulation exceeds the optionally specified maximum time, then the simulation is finished. Else all notifications at this time are performed and removed. Processes sensitive on these notifications are made runnable. If at least one runnable process exists at the end of this phase, simulation continues with step 2. Else the simulation is finished.

[1] If the same implementation of the simulation kernel is used to simulate the same SystemC program with the same inputs, then the process order shall remain the same.

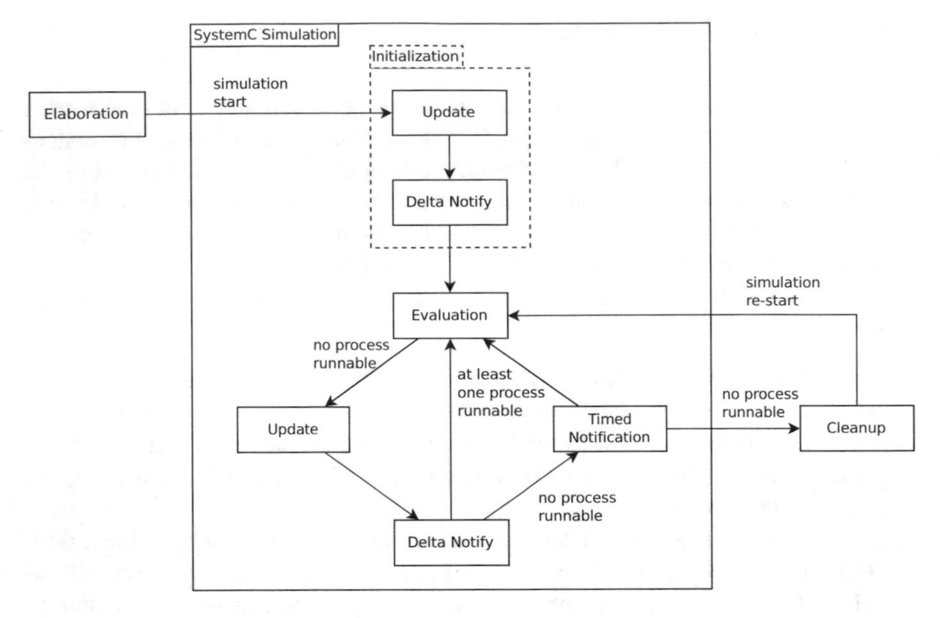

Fig. 2.1 Execution phases of a SystemC program (from [85])

After simulation the remaining statements after *sc_start* will be executed. Optionally *sc_start* can be called again, thus resuming the simulation. In this case the *initialization* phase will not be called. The simulation will directly continue with the *evaluation* phase. An overview of the different phases and transitions between them is shown in Fig. 2.1.

The non-determinism in the *evaluation* phase of the simulation, due to multiple process scheduling alternatives, is one of the reasons that give rise to the state explosion problem in verification of SystemC designs. Thus, it is necessary to consider all possible process execution orders, in addition to all possible inputs, to avoid missing relevant behavior. Symbolic simulation, that is presented later in Chap. 4, enables to do so very efficiently.

For more information on SystemC and TLM please refer to [19, 71, 74] as well as the official standard reference [113, 163].

2.2 RISC-V

RISC-V is an open and free instruction set architecture (ISA) that gained enormous momentum in recent years. Many and diverse application areas are available for RISC-V ranging from IoT and automotive to production and multi-media. For example, Western Digital has announced to produce over one billion RISC-V cores per year and announced an own open-source RISC-V core called *SweRV* [17, 220].

Another example is Nvidia, they are also working on integrating RISC-V cores into their designs [167, 178]. In addition, several other large companies such as Google, Samsung, and NXP are Platinum level members of the RISC-V foundation [151].

2.2.1 *ISA Overview*

The RISC-V ISA itself is designed in a very scalable and modular way. The ISA consists of a mandatory base integer instruction set and various optional extensions. The integer set is available in three different configurations with 32, 64, and 128 bit width registers, respectively: RV32I, RV64I, and RV128I. Additionally, the RV32E configuration, which is essentially a lightweight RV32I with a reduced number of registers, is available and intended for (very) small embedded devices. The base Integer set consists of a small set of instruction that enable to perform arithmetic and logic computation, signed and unsigned comparison, branches and jumps, special register access, load and store operations as well as loading of immediate values and register transfer operations. In total the Integer set has 47 instructions. Instruction set extensions are denoted with a single letter, e.g. M (integer multiplication and division), A (atomic instructions), C (compressed instructions), F (single-precision floating-point), D (double-precision floating-point), etc. The RV32IM ISA denotes a core with 32 bit registers implementing the base Integer instruction set and the Multiplication extension. The instruction set combination IMAFD is also abbreviated as G and denotes the standard RISC-V instruction set. A comprehensive description of the RISC-V instruction set is available in the specification [218].

In addition to the instruction sets, the RISC-V ISA also comprises a privileged architecture description which is available in the second volume of the RISC-V ISA specification [219]. It defines *Control and Status Registers* (CSRs), which are registers serving a special purpose. For example the *MISA* (Machine ISA) register is a read-only CSR that contains information about the supported ISA. Another example is the *MTVEC* (Machine Trap-Vector Base-Address) CSR that stores the address of the trap/interrupt handler. In addition, the privileged architecture description provides a small set of instructions for interrupt handling (*WFI*, *MRET*) and interacting with the system environment (*ECALL*, *EBREAK*).

As an example, the Atomic instruction set extension is detailed in the following, since it is referred to again in Chap. 3.

2.2.2 *Atomic Instruction Set Extension*

The RISC-V Atomic instruction set extension enables implementation of synchronization primitives between multiple RISC-V cores. It provides two types of atomic instructions:

(1) A set of *AMO* instructions. For example, *AMOADD.W* loads a word X from memory, performs an addition to X, and stores the result back into the same memory location. Furthermore, X is also stored in the destination register RD of the *AMO* instruction. The load and store operation of the *AMO* instructions are executed in an atomic way, i.e. they are not interrupted by other memory access operations from other cores.
(2) The LR (Load Reservation)/SC (Store Conditional) instructions. LR loads a word from memory and places a reservation on the load address. The reservation size must at least include the size of the memory access. SC stores a result to memory. SC succeeds if a reservation exists on the store address and no other operation has invalidated the reservation, e.g. a store on the reserved address range. Otherwise, SC fails with an exception, which in turn triggers a trap. A sequence of corresponding LR/SC instruction should satisfy the RISC-V *"forward-progress"* property, i.e. essentially, contain no more than 16 simple integer instructions (i.e. no loads, stores, jumps, or system calls) between the LR and SC pair.

2.3 Coverage-Guided Fuzzing

Fuzzing is a very effective SW testing technique. More recent fuzzing approaches in the SW domain employ the so-called *mutation* based technique. It mutates randomly created data and is guided by code coverage. Notable representatives in this *Coverage-guided Fuzzing* (CGF) category are the LLVM-based *libFuzzer* [141] and *AFL* [9], which both have been shown very effective and revealed various new bugs [9, 141]. In the following libFuzzer is presented in more detail, since libFuzzer is used in this book.

2.3.1 LibFuzzer Core

LibFuzzer aims to create inputs (i.e. each input corresponds to a testcase and is simply a binary bytestream) in order to maximize the code coverage of the *Software Under Test* (SUT). Therefore, the SUT is instrumented by the *Clang* compiler to report coverage information that is recognized by libFuzzer. Input data is transformed by applying a set of pre-defined mutations (shuffle bytes, insert bit, etc.) randomly. The mutations are guided by the observed code coverage. For example, inputs as well as mutation sequences that result in new code coverage are ranked with a higher score to increase the probability of re-using them to generate new input. The input size is periodically increased.

Technically libFuzzer is linked with the SUT into a single executable, hence performs the so-called *in-process* fuzzing, and allows to pass inputs to the SUT as well as receive coverage information back through specific interface functions. The

```
1 extern "C" int LLVMFuzzerTestOneInput(const uint8_t *Data, size_t Size) {
2   // allows to run some initialization code once, first time this function is called
3     static bool initialized = do_initialization();
4
5     // ... process the input ...
6
7   // tell libFuzzer that the input has been processed
8     return 0;
9 }
```

Fig. 2.2 Interface function [141] that the SUT provides and that is repeatedly called by libFuzzer during the fuzzing process

fuzzing process itself works as follows: libFuzzer provides the C++ main function and hence is the entry point of the execution. After a short initialization, libFuzzer starts to repeatedly call the input interface function of the SUT for each generated input. Figure 2.2 shows the signature of the input interface function. It receives a data pointer and data size which represents the input (i.e. single testcase). During execution of the input in the SUT the coverage interface functions provided by libFuzzer will be called to collect coverage information for this run. These function calls are automatically instrumented into the SUT by compiling it with Clang (and using appropriate compilation options).

2.3.2 LibFuzzer Extensions

To further guide the fuzzing process and enable domain specific optimization, libFuzzer provides a simple interface to integrate custom mutation procedures into the fuzzing process. Figure 2.3 shows the signature of the interface function. When the SUT simply provides this function, it will automatically be called by libFuzzer every time an input is mutated. It receives four arguments: the current input represented as Data pointer and Size as well as a new maximum size (MaxSize) of the result input and a Seed to apply deterministic pseudo-random mutations. Please note, the current input is mutated in place by directly modifying the Data pointer. The result of the function is the size of the new input (which must be less than or equal to MaxSize). It is possible to access the libFuzzer provided mutators by calling the *LLVMFuzzerMutate* function as also shown in Fig. 2.3.

Fuzzing is often combined with sanitizers and other automatic runtime checks to check for additional error sources during testcase execution. A sanitizer instruments the source code during compilation to automatically add additional behavior. Besides the coverage sanitizer, that instruments the source code to emit coverage information for libFuzzer, Clang also provides sanitizers to check for various memory access errors as well as undefined behavior.

```
1 extern "C" size_t LLVMFuzzerCustomMutator(uint8_t *Data, size_t Size,
2                                           size_t MaxSize, unsigned int Seed) {
3   // It is possible to add a custom data mutator here.
4   // ...
5
6   // In this case simply call the existing libFuzzer mutator function that randomly
        selects an existing libFuzzer mutator.
7   return LLVMFuzzerMutate(Data, Size, MaxSize);
8 }
```

Fig. 2.3 Interface function [141] that the SUT can provide to implement custom mutators

2.4 Symbolic Execution

Symbolic execution [27, 122] is a formal program analysis technique to efficiently and exhaustively explore the whole state space by using symbolic expressions and constraints. Symbolic execution as described in the following is a standard technique and often serves as foundation for more advanced state space exploration techniques that leverage symbolic execution and integrate additional optimizations and complementary state space reduction techniques. Symbolic simulation that we present in Chap. 4 for formal SystemC verification is such an example.

2.4.1 *Overview*

Symbolic execution analyzes the behavior of a program pathwise by treating inputs as symbolic expression. A symbolic expression represents a set of concrete values and is constrained by the *Path Condition* (PC). Symbolic execution updates the symbolic execution state, which consists of symbolic expressions and the PC, according to the execution semantic of each instruction alongside the execution path.

In case of a branch instruction `if` C `then` $B1$ `else` $B2$ with symbolic branch condition C the symbolic execution state s is split into two independent execution states s_T and s_F, respectively. The new states s_T and s_F represent the execution path that continues at the if-branch $B1$ and else-branch $B2$, respectively. The PC for each state is updated accordingly as $PC(s_T) := PC(s) \wedge C$ and $PC(s_F) := PC(s) \wedge \neg C$. An SMT solver is used to determine if s_T and s_F are feasible, i.e. their PC is satisfiable. Only feasible paths will be explored further. Loops are handled similarly to branches based on the loop condition. Thus, symbolic execution effectively creates a tree of execution paths where the PC represents the constraints under which specific position in the program will be reached.

For verification purposes, symbolic execution provides the *make_symbolic*, *assume* and *assert* functions: *assume(C)* adds C to the current PC to prune irrelevant paths and *assert(C)* calls an SMT solver to check for assertion violation, i.e. this is the case iff $PC \wedge \neg C$ is satisfiable. Besides assertion violation, symbolic

```
1  int a = 0;
2  int b = 0;
3
4  int main() {
5    make_symbolic(&a, sizeof(a));
6
7    if (a >= 0)
8      b = a + 1;
9    else
10     b = -a;
11
12   assert (b >= 0);
13   return 0;
14 }
```

Fig. 2.4 Example program to illustrate symbolic execution

execution can also be extended to check for other generic error sources such as memory out of bounds access and division by zero. In case of an error, a counterexample with concrete values for all symbolic input expressions is provided to reproduce the error. The *make_symbolic* function allows to make a variable symbolic, i.e. assign an unconstrained symbolic expression to that variable.

2.4.2 Example

For illustration Fig. 2.4 shows an example. It is a simple program with two global variables a and b. It starts by making a symbolic in Line 5. Then, it executes an *if-then-else* block (Lines 7–10). Finally, it validates the value of variable b in Line 12.

The symbolic execution state of the program consists of the values of the two variables a and b as well as the PC (and also the current position in the program). Figure 2.5 shows the symbolic execution tree for this example. Each node represents an update of the symbolic execution state and is annotated with the line number of the instruction which resulted in that update. Due to the branch in Line 7 the execution is split into two paths that continue at Lines 8 and 10, respectively. In one case the assertion is satisfied, for the other path the assertion is violated due to an integer overflow in case of $a = INT\_MAX$.

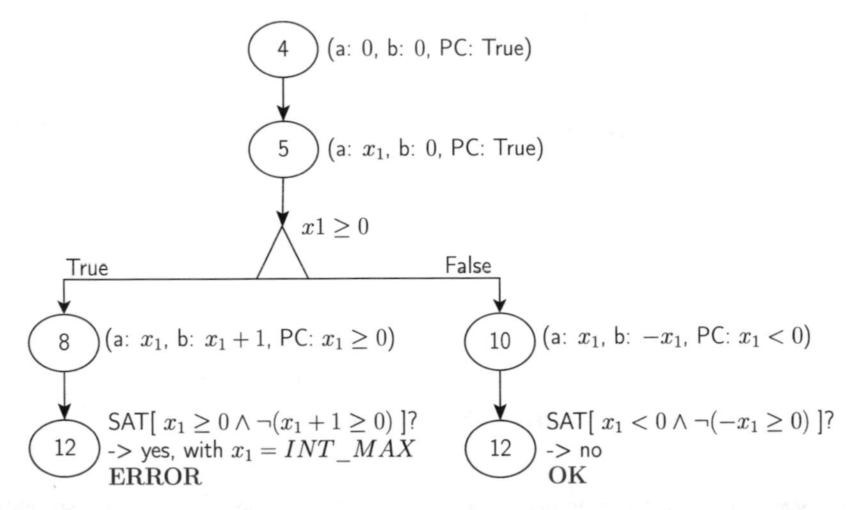

Fig. 2.5 Symbolic execution tree for the example shown in Fig. 2.4. Nodes correspond to line numbers and are annotated with the symbolic execution state after executing that line. Node Line 12 is annotated with the symbolic expression SMT solver query to decide if the assertion is violated in Line 12

Chapter 3
An Open-Source RISC-V Evaluation Platform

Enormous innovations are enabled by today's embedded systems, in particular the *Internet-of-Things* (IoT) applications where every device is connected to the Internet. Forecasts see additional economic impact resulting from industrial IoT. In the last years the complexity of IoT devices has been increasing steadily with various conflicting requirements. On the one hand, IoT devices need to provide smart functions with a high performance including real-time computing capabilities, connectivity, and remote access as well as safety, security, and high reliability. On the other hand, they have to be cheap, work efficiently with an extremely small amount of memory and limited resources and should further consume only a minimal amount of power to ensure a very long lifetime.

To meet the requirements of a specific IoT system, a crucial component is the processor. Stimulated from the enormous momentum of open-source software, a counterpart on the hardware side recently emerged: *RISC-V* [218, 219]. RISC-V is an open-source *Instruction Set Architecture* (ISA), which is license-free and royalty-free. The ISA standard is maintained by the non-profit RISC-V foundation and is appropriate for all levels of computing systems, i.e., from micro-controllers to supercomputers. The RISC-V ecosystem is rapidly growing, ranging from HW, e.g., various HW implementations (free as well as commercial) to high-speed *Instruction Set Simulators* (ISSs) These ISSs facilitate functional verification of RTL implementations as well as early SW development to some extent. However, being designed predominantly for speed, they can hardly be extended to support further system-level use cases such as design space exploration, power/timing/performance validation, or analysis of complex HW/SW interactions.

A major industry-proven approach to deal with these use cases in earlier phases of the design flow is to employ *Virtual Prototypes* (VPs) [139] at the abstraction of *Electronic System Level* (ESL) [12]. In industrial practice, the standardized C++-based modeling language SystemC and *Transaction Level Modeling* (TLM) techniques [71, 113] are being heavily used together to create VPs. Depending on the specific use case, advanced state-of-the-art SystemC-based techniques beyond

V. Herdt et al., *Enhanced Virtual Prototyping*,
https://doi.org/10.1007/978-3-030-54828-5_3

functional modeling (see e.g., [76, 80, 89, 91, 162, 169, 200]) are to be applied on top of the basic VPs. The much earlier availability and the significantly faster simulation speed in comparison to RTL are among the main benefits of SystemC-based VPs.

First, in Sect. 3.1 we propose a RISC-V-based VP to further expand and bring the benefits of VPs to the RISC-V ecosystem. With the goal of filling the mentioned gap in supporting further system-level use cases, SystemC is necessarily the language of choice. The VP is therefore implemented in standard-compliant SystemC and TLM-2.0 and designed as extensible and configurable platform with a generic bus system. We provide a 32- and 64-bit RISC-V core supporting the RISC-V standard instruction set with different privilege levels, the RISC-V-specific CLINT, and PLIC interrupt controllers and an essential set of peripherals. Furthermore, we support simulation of (mixed 32- and 64-bit) multi-core platforms, provide SW debug and coverage measurement capabilities, and support the FreeRTOS and Zephyr operating systems. We further demonstrate the extensibility and configurability of our VP by three examples: addition of a sensor peripheral, describing the integration of the GDB SW debug extension, and configuration to match the RISC-V HiFive1 board from SiFive. Our evaluation demonstrates the quality and applicability of our VP to real-world embedded applications and shows the high simulation performance of our VP. Finally, our RISC-V VP is fully open source[1] (MIT license) to stimulate further research and development. The approach has been published in [95, 96, 106].

Then, in Sect. 3.2 we propose an efficient core timing model and integrate it into the RISC-V VP core, to enable fast and accurate performance evaluation for RISC-V-based systems. Our timing model is attached to the core using a set of well-defined interfaces that decouple the functional from the non-functional aspects and enable easy customization of the timing model. This is very important, because the timing strongly depends on the actual system, e.g., the structure of the execution pipeline in the CPU core or the use of caches. Our interface and timing model allows to consider pipelining, branch prediction, and caching effects. This feature set makes our core timing model suitable for a large set of embedded systems. Our timing model essentially works by observing the executed instructions and memory access operations at runtime. A previous static analysis of the binary or access to the source code is not required. As a case study, we provide a timing configuration matching the RISC-V HiFive1 board from SiFive. The HiFive1 core combines a five-level execution pipeline with an instruction cache and a branch prediction unit. Our experiments demonstrate that our approach allows to obtain very accurate performance evaluation results (less than 5% mismatch on average compared against a real HW board) while still retaining a high simulation performance (less than 20% performance overhead on average for the VP simulation). We use the *Embench* benchmark suite for evaluation, which is a modern benchmark suite tailored for embedded systems. Our core timing model including the integration with the open-source RISCV VP is available as open source too in order to further

[1]Available at https://github.com/agra-uni-bremen/riscv-vp, and for more information and most recent updates also visit our site at www.systemc-verification.org/riscv-vp.

advance the RISC-V ecosystem and stimulate research. The approach has been published in [104].

3.1 RISC-V-Based Virtual Prototype

The RISC-V ecosystem already has various high-speed ISSs such as the reference simulator Spike [197], RISCV-QEMU [185], RV8 [186], or DBT-RISE [42]. They are mainly designed to simulate as fast as possible and predominantly employ dynamic binary translation (to x86_64) techniques. This is however a trade-off as accurately modeling power or timing information for instructions becomes much more challenging.

The full-system simulator gem5 [18], at the time of writing also has initial support for RISC-V. gem5 provides more detailed models of processors and memories and can in principle also be extended for accurate modeling of extra-functional properties. Renode [176] is another full-system simulator with RISC-V support. Renode puts a particular focus on modeling and debugging multi-node networks of embedded systems. However, both gem5 and Renode employ a different modeling style and thus hinder the integration of advanced SystemC-based techniques.

FORVIS [58] and GRIFT [67] are Haskell-based implementations that aim to provide an executable formalization of the RISC-V ISA to be used as foundation for several (formal) analysis techniques. SAIL-RISCV [184] aims to be another RISC-V formalization that is implemented in Sail, which is a special language for describing ISAs with support for generation of simulator back-ends (in C and OCaml) as well as theorem-prover definitions.

The project SoCRocket [189] that develops an open-source SystemC-based VP for the SPARC V8 architecture can be considered comparable to our effort.

Finally, commercial VP tools such as Synopsys Virtualizer or Mentor Vista might also support RISC-V but their implementation is proprietary.

In the following we present our open-source RISC-V based VP that is implemented in SystemC TLM and attempts to close the gap on virtual prototyping in the RISC-V ecosystem.

3.1.1 RISC-V-Based VP Architecture

The VP is implemented in SystemC and designed as extensible and configurable platform around a RISC-V RV32/64IMAC (multi-)core with a generic bus system employing TLM 2.0 communication and support for the GNU toolchain with SW coverage measurement (GCOV) and debug capabilities (GDB). Overall, the VP consists of around 12,000 lines of C++ code with all extensions. Figure 3.1 shows an overview of the VP architecture. In the following we present more details.

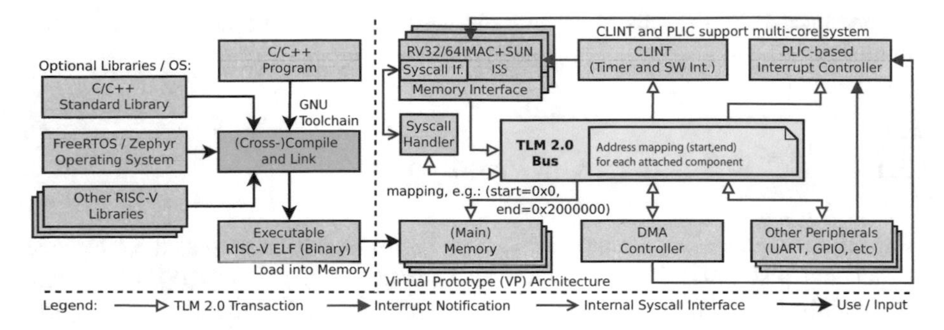

Fig. 3.1 Virtual prototype architecture overview

3.1.1.1 RV32/64 (Multi-)Core

The CPU core loads, decodes, and executes one instruction after another. RISC-V compressed instructions are expanded on the fly into regular instructions in a pre-processing step before being passed to the normal execution unit. We provide a 32-bit and 64-bit core supporting the RISC-V RV32IMAC and RV64IMAC instruction set, respectively. Besides the mandatory Machine mode, each core implements the RISC-V Supervisor and User mode privilege levels and provides support for user mode interrupt and trap handling (N extension). This includes the *Control and Status Register* (CSR) for the corresponding privilege levels (as specified in the RISC-V privileged architecture specification [219]) as well as instructions for interrupt handling (*wfi*, *m/s/uret*) and environment interaction (*ecall*, *ebreak*). We will provide more details on the implementation of interrupt handling and system calls (environment interaction) in the following sections.

Multiple RISC-V cores can be integrated to build a multi-core platform. It is also possible to mix 32- and 64-bit cores. The Atomic extension provides instructions to enable synchronization of the cores. Each core is attached to the bus through a memory interface. Essentially, the memory interface translates load/store requests into TLM transactions and ensures that the atomic instructions are handled correctly. We will provide more details on simulation of multi-core platforms in Sect. 3.1.4.

3.1.1.2 TLM 2.0 Bus

The TLM bus is responsible for routing transactions from an initiator, i.e., (bus) master, to a target. Therefore, all target components are attached to the TLM bus at specific non-overlapping address ranges. The bus will match the transaction address with the address ranges and dispatch the transaction accordingly to the matching target. Please note, in this process, the bus performs a *global-to-local* address translation in the transaction. For example, assume that a sensor component is mapped to the address range (start=0x50000000, end=0x50001000) and the transaction address is 0x50000010, then the bus will route the transaction to the sensor and change the transaction address to 0x00000010 before passing it on to the sensor. Thus the sensor works on local address ranges. The TLM bus supports multiple masters initiating transactions. Currently, the CPU core as well as the DMA controller are configured as bus masters. Please note that a single component can be both master and target, as for example the DMA controller receives transactions initiated by the CPU core to configure the source and destination address ranges and also initiates transactions by itself to perform the memory access operations without the CPU core.

3.1.1.3 Traps and Interrupts

Both traps and interrupts result in the CPU core performing a context switch to the trap/interrupt handler (based on a SW configurable address stored in the *MTVEC* CSR). Traps are raised to perform a system call or when an execution exception, e.g., invalid memory access, is encountered.

Two sources of interrupts are available: (1) local and (2) external. Essentially, there are two sources of local interrupts: SW as well as timer interrupts generated by the RISC-V-specific CLINT (Core Local INTerruptor). The timer is part of the CLINT, and the interrupt frequency can be configured for each core through memory mapped I/O. CLINT also provides a memory mapped register for each core to trigger a SW interrupt for the corresponding core. External interrupts are all remaining interrupts triggered by the various components in the system. To handle external interrupts, we provide the RISC-V-specific PLIC (Platform Level Interrupt Controller). PLIC will collect and prioritize all external interrupts and then route them to each CPU core one by one. The core that claims the interrupt first will process it. According to the RISC-V specification, external interrupts are processed with higher priority than local interrupts, and SW interrupts are higher prioritized than timer interrupts. We will describe the interrupt handling process in more detail in Sect. 3.1.2.

3.1.1.4 System Calls

The C/C++ library defines a set of system calls as abstraction from the actual execution environment. For example, the *printf* function performs the formatting in platform independent C code and finally invokes the *write* system call with a fixed char array. Typically, an embedded system provides a trap handler that redirects the *write* system call to a UART/terminal component.

We also provide a *SyscallHandler* component to emulate system calls of the C/C++ library by redirecting them to the simulation host system. Our emulation layer for example allows us to open and read/write from/to files of the host system. We use this functionality for example to support SW coverage measurement with GCOV.

The syscall handler can be called in one of two ways: (1) through a trap handler that redirects the system call to the syscall handler from SW using memory mapped I/O (this approach enables a flexible redirection of selected system calls), or (2) directly intercept the system call (i.e., the RISC-V *ECALL* instruction) in the CPU core, instead of jumping to the trap handler. The behavior is configurable per core. We will describe the system call handling process in more detail in Sect. 3.1.2.

3.1.1.5 VP Initialization

The main function in the VP is responsible to instantiate, initialize, and connect all components, i.e., set up the architecture. An ELF loader is provided to parse and load an executable RISC-V ELF file into the memory and set up the program counter in the CPU core accordingly. Finally, the SystemC simulation is started. The ELF file is produced by the GNU toolchain by (cross-)compiling the application program and optionally linking it with the C/C++ standard library or other RISC-V libraries. We also support a bare-metal execution environment without any additional libraries and test our VP to work with the FreeRTOS and Zephyr operating systems.

3.1.1.6 Timing Model

We provide a simple and configurable instruction-based timing model in the core, and by following the TLM 2.0 communication standard, transactions can be annotated with optional timing information to obtain a more accurate timing model of the executed software. We also provide a plugin interface to integrate external timing information into the core, and we are working on integrating more accurate timing models.

3.1.2 VP Interaction with SW and Environment

In this section we present more details on the HW/SW interaction, in particular on interrupt handling, and environment interaction via system calls in our VP.

3.1.2.1 Interrupt Handling and HW/SW Interaction

In the following we present an example application that periodically accesses a sensor to demonstrate the interaction between hardware (VP-side) and software with a particular focus on interrupt handling. We first describe the software application running on the VP and then present a minimal assembler bootstrap code to initialize interrupt handling and describe how interrupts are processed in more detail. Later

```
1  #include "stdint.h"
2  #include "irq.h"
3
4  static volatile char * const TERMINAL_ADDR = (char * const)0x20000000;
5  static volatile char * const SENSOR_INPUT_ADDR = (char * const)0x50000000;
6  static volatile uint32_t * const SENSOR_SCALER_REG_ADDR = (uint32_t *
       const)0x50000080;
7  static volatile uint32_t * const SENSOR_FILTER_REG_ADDR = (uint32_t *
       const)0x50000084;
8
9  _Bool has_sensor_data = 0;
10
11 void sensor_irq_handler() {
12   has_sensor_data = 1;
13 }
14
15 void dump_sensor_data() {
16   while (!has_sensor_data) {
17     asm volatile ("wfi");
18   }
19   has_sensor_data = 0;
20
21   for (int i=0; i<64; ++i) {
22     *TERMINAL_ADDR = *(SENSOR_INPUT_ADDR + i);
23   }
24 }
25
26 int main() {
27   register_interrupt_handler(2, sensor_irq_handler);
28
29   *SENSOR_SCALER_REG_ADDR = 5;
30   *SENSOR_FILTER_REG_ADDR = 2;
31
32   for (int i=0; i<3; ++i)
33     dump_sensor_data();
34
35   return 0;
36 }
```

Fig. 3.2 Example application running on the VP to demonstrate the hardware/software interaction

in Sect. 3.1.5.1 we present the corresponding SystemC-based sensor implementation in our VP.

Software Side Figure 3.2 shows an example application that reads data from a sensor and copies the data to a terminal component. The sensor and terminal are accessed through memory mapped I/O. Their addresses are defined at the top of the program. They need to match with the configuration in the VP. The sensor periodically triggers an interrupt, denoting that new data is available. The main function starts by registering an interrupt handler for the sensor interrupt (Line 27). Again, the interrupt number specified in SW has to match the configuration in the VP. Next, the sensor is configured in Lines 29–30 using memory mapped I/O. The

```
1  .globl _start
2  .globl main
3  .globl level_1_interrupt_handler
4
5  _start:
6  la t0, level_0_interrupt_handler
7  csrw mtvec, t0 # register interrupt/trap handler
8  csrsi mstatus, 0x8 # enable interrupts in general
9  li t1, 0x888
10 csrw mie, t1 # enable external/timer/SW interrupts
11 jal main
12
13 # stop simulation with the *exit* system call
14 li a7, 93 # syscall exit has number 93
15 li a0, 0 # argument to exit
16 ecall # RISC-V system call
17
18 level_0_interrupt_handler:
19 # ... store registers on the stack if necessary ...
20 csrr a0, mcause
21 jal level_1_interrupt_handler
22 # ... re-store registers in case they have been saved ...
23 mret # return from interrupt/trap handler
```

Fig. 3.3 Bare-metal bootstrap code demonstrating interrupt handling

scaler denotes how fast sensor data is generated and the filter setting what kind of post-processing is performed on the data. Finally, the copy process is iterated for three times (Lines 32–33) before the program terminates. Each iteration starts by waiting for sensor data (Lines 16–18). The global Boolean flag *has_sensor_data* is used for synchronization. It is set in the interrupt handler (Line 12) and unset again immediately after the waiting loop (Line 19). Please note that the *wfi* instruction will power down the CPU core until the next interrupt occurs.

Bootstrap Code and Interrupt Handling Figure 3.3 shows the essential parts of a bare-metal boostrap code, which is written in assembler and linked with the application code, to handle interrupts.[2] The *_start* label is the entry point of the whole program. The registers *mtvec*, *mstatus*, *mie*, and *mcause* are CSRs that essentially store the interrupt handler address, core status information, enabled interrupts, and interrupt source, respectively. The instructions *csrr* and *csrw* read and write a CSR into and from a normal CPU register, respectively. The instruction *csrsi* sets the bits in the CSR based on the provided immediate value. Before the main function is called (Line 11), the interrupt handler base address (level-0) is stored in *mtvec* (Lines 6–7) and all interrupts are enabled (Lines 8–10). After the

[2]Support for integration with the C/C++ library is also available, e.g., by executing the instructions at the beginning of the main function or integrating them directly into the *crt0.S* file, which is the entry point of the C library and similarly to the bare-metal code also calls the main function after performing some basic initialization tasks.

main function returns, the exit system call is invoked to terminate the VP simulation (Lines 14–16). We provide more details on system calls in the next section.

In general, an interrupt can occur at any time during execution of the application SW. All interrupts propagate to the PLIC (i.e., the RISC-V interrupt controller) first and are prioritized there. The CPU core only receives a notification that some interrupt is pending and needs to be processed. The core will prepare the interrupt by storing the program counter into the *mepc* CSR and setting the *mcause* CSR appropriately (to denote an interrupt in this case). The core then reads the base address from the *mtvec* CSR and sets the program counter to that address, i.e., effectively directly jumping to the level-0 interrupt handler (first instruction at Line 20). The interrupt handler (level-0) first in Line 20 reads the reason (i.e., local or external interrupt) for the interrupt into the *a0* CPU register, which according to the RISC-V calling convention [179] stores the first argument of a function call. Then in Line 21 an interrupt handler implemented in C is called (level-1, not shown in this example). Essentially, this level-1 handler deals with a local timer interrupt by resetting the timer with an external interrupt by asking the IC for the actual interrupt number with the currently highest priority (through a memory mapped register access) and then calls the application-provided interrupt handler function (Lines 11–13 in Fig. 3.2, this step is ignored if none has been registered for the interrupt number). Finally, the *mret* instruction restores the previous program counter from the *mepc* CSR. Please note that the level-0 handler typically stores and re-stores the register values by pushing and popping them to/from the stack before/after calling the level-1 handler, respectively.

3.1.2.2 Environment Interaction: Syscall Emulation and C/C++ Library

We provide an emulation layer for executing system calls by redirecting them to the host system running the VP simulation. This requires passing arguments from the guest application into the host system and integrating the return values back into the guest application (i.e., memory of the VP). Implementing syscalls enables support for the C/C++ standard library. Furthermore, we can directly use GCOV to track the coverage of the applications simulated on our VP (the GCOV instrumentation requires syscall support to open and write to files).

For example, consider the *printf* function provided by the C standard library. Most of its functionality is implemented as portable C code independent of the execution environment. Essentially, the *printf* function will apply all formatting rules and create a simple char buffer, which is then passed to the *write* system call. At this point, interaction with the execution environment is required. Figure 3.4 shows the relevant part of a stub that is provided in the RISC-V port of the C library. Essentially, the arguments of the system call are stored in the CPU registers *a0–a3* and the syscall number in *a7*. Then the *ecall* instruction is executed. The VP simulator will detect the *ecall* instruction and directly execute the syscall on the host

```
1  #define SYS_write 64
2
3  ssize_t write(int fd, const void *buf, size_t count) {
4    return syscall(SYS_write, fd, (long)buf, count, 0);
5  }
6
7  long syscall(long n, long _a0, long _a1, long _a2, long _a3) {
8    // store arguments in CPU register and trigger ecall
9    register long a0 asm("a0") = _a0;
10   register long a1 asm("a1") = _a1;
11   register long a2 asm("a2") = _a2;
12   register long a3 asm("a3") = _a3;
13   register long a7 asm("a7") = n;
14
15   // special RISC-V instruction denoting a system call
16   asm volatile ("ecall" : "+r"(a0) : "r"(a1), "r"(a2), "r"(a3), "r"(a7));
17
18   // store potential error code and return result
19   if (a0 < 0) {
20     errno = -a0;
21     return -1;
22   } else {
23     return a0;
24   }
25 }
```

Fig. 3.4 System call handling stub linked with the C library (guest side, executed on the VP host system). This example listing is based on the RISC-V *newlib* port https://github.com/riscv/riscv-newlib

system as shown in Fig. 3.5.[3] In case of the *write* syscall, a pointer argument *buf* is passed. This is a pointer value from the guest system, i.e., an index in the VP byte memory array *mem*, and has to be translated to a host memory pointer in order to execute the *write* syscall on the host system. Therefore, the guest_to_host_pointer function (Line 5) adds the base address of the VP byte memory array, i.e., *mem* + *buf*. The result of the syscall is stored in the *a0* register and passed back to the C library. We have implemented other syscalls in a similar way to the *write* syscall.

In general the guest and host system have a different architecture with different word sizes, e.g., in our case the guest system (which is simulated in the VP) can be a 32-bit and the host system (which runs the VP) is a 64-bit system. Therefore, one has to be careful when data is passed between the guest and the host. Primitive types, e.g., int and bool, can be passed directly from the guest to the host, because our host system running the VP uses data types with equal or larger sizes, thus no information is lost when passing the arguments. When passing values back from the host, a check can be performed, if necessary, to ensure that no relevant information

[3]It is also possible to execute a trap handler, similar to the interrupt handler described in the previous section (e.g., essentially, jump to the level-0 interrupt handler with the *mcause* CSR being set to the syscall identifier) and then redirect the write to, e.g., a UART/Terminal component.

```
1  #define SYS_write 64
2
3  // execute syscall on the host system (SyscallHandler)
4  ssize_t sys_write(int fd, const void *buf, size_t count) {
5     const void *p = (const void *)guest_to_host_pointer(buf);
6     return write(fd, p, count);
7  }
8
9  long execute_syscall(long n, long _a0, long _a1, long _a2, long _a3) {
10    switch (n) {
11       case SYS_write:
12          return sys_write(_a0, (const void *)_a1, _a2);
13       //...
14    }
15 }
16
17 // function inside the CPU core
18 void execute_step() {
19   auto instr = mem_if->load_instr(program_counter);
20   auto op = decode(instr);
21
22   switch (op) {
23     case Opcode::ECALL: {
24       if (intercept_syscalls_option) {
25         // intercept and redirect syscall to host system
26         regs[a0] = execute_syscall(regs[a7], regs[a0], regs[a1], regs[a2], regs[a3]);
27       } else {
28         // jump to SW trap handler (let SW decide how to handle syscall)
29         // SW will either direct the syscall to a peripheral or our SyscallHandler
              component (which is what the core does directly when intercepting
              syscalls)
30         raise_trap(EXC_ECALL);
31       }
32     } break;
33     //...
34   }
35 }
```

Fig. 3.5 Concept of system call execution on the VP, either redirects to the host system or takes trap

is truncated, e.g., due to casting a 64-bit value into a 32-bit one. Pointer arguments need to be translated to host addresses, as described above, before accessing them on the host system. A write access is thus directly propagated back to the guest application. Structs can be accessed and copied recursively, considering the rules for accessing primitive and pointer types.

3.1.3 VP Performance Optimizations

In this section we discuss two performance optimizations for our VP that result in significant simulation speed-ups. The first optimization is a direct memory interface to fetch instructions and perform load/store operations from/to the (main) memory more efficiently. The second is a temporal decoupling technique with local time quantums to reduce the number of costly context switches, especially, in the CPU core simulation. We describe both techniques in the following sections.

3.1.3.1 Direct Memory Interface

The CPU core translates every load and store operation into a transaction, which is routed through the bus to the target. Most of the time the main memory is the target of the access. Always accessing the memory through a bus transaction can be very costly. Even more so, because fetching the next instruction requires to load it from the memory too. Thus, at least one bus transaction is executed for every instruction. To optimize the access of the main memory and in particular instruction fetching, we provide two proxy classes with a direct memory interface. The direct memory interface stores the address offset where the memory is mapped in the overall address space as well as the size and pointer to the start of the memory. We have a proxy class for fetching instructions and one to access the memory in general, i.e., to perform load/store byte/half/word instructions. With the proxy classes enabled, the CPU core will first query the proxy class. It will match in case the main memory is accessed (for the instruction proxy class, we only allow to fetch instructions from main memory) and otherwise convert the access into a transaction and normally route it through the bus.

3.1.3.2 Local Time Quantums

A SystemC-based simulation is orchestrated by the SystemC simulation kernel that switches execution between the various threads. While this is not a performance problem for most components, since they become runnable on very specific events, context switching can become a major bottleneck in simulating the CPU core. The reason is that a direct implementation will perform a context switch after executing every instruction, because simulation time has passed and the SystemC kernel needs to check for other runnable threads to perform synchronization. However, most of the time no other thread becomes runnable and the CPU thread is resumed again. Even if some other thread would become runnable, it is still fine to keep running the CPU thread for some time (ahead of the global simulation time of the system). For example, even if the sensor thread would be runnable and would trigger an interrupt once executed, delaying the sensor thread execution for a small amount of time and keeping the CPU thread running should not have influence on the functional behavior of the system. In general the software does have any knowledge of the exact timing behavior and thus is written in such a way, e.g., by employing locks and flags, to always wait for certain conditions.

3.1.4 *Simulation of Multi-Core Platforms*

Our VP provides support for simulation of RISC-V multi-core platforms. We support to instantiate and mix multiple 32- and 64-bit cores. Each core is attached to the bus using a (local TLM) memory interface. Furthermore, each core is assigned a unique identifier, starting with zero, during instantiation. This core identifier is denoted as *hart-id*. Access to the hart-id is provided through the read-only SW accessible *mhartid* CSR. Based on the hart-id, the SW can control the execution for each core. Furthermore, the SW is using atomic instructions for synchronization.

In the following we show an example RISC-V multi-core SW for illustration and provide more details on the implementation of the RISC-V atomic ISA extension.

3.1.4.1 Example Bare-Metal Multi-Core SW

Figure 3.6 shows an example bare-metal SW that demonstrates the multi-core simulation concept from the SW side. For illustration purposes, we assume that the platform consists of two cores. Both cores start at the same time and use the same entry point (Line 5, i.e., the *_start* label). By reading the *mhartid* CSR, the SW obtains the id, either 0 or 1, of the executing core (Line 6). The core id is stored in register a0. Based on the id, the SW control flow is manipulated. Each core is initializing its stack pointer (sp) address to a separate area (Lines 13 and 16). Finally, an external main function is called (Line 20) with the core id passed as first argument (according to the RISC-V calling convention, the first argument is provided in register a0). The code after the main function ensures that only the last core can proceed to the exit system call (Lines 32–34) and stop the simulation. The first core that returns from the main function keeps spinning in the loop at Line 29. This synchronization mechanism is achieved by using an AMO instruction to read and increment a shared counter.

3.1.4.2 Implementation of the Atomic ISA Extension

Figure 3.7 shows the relevant part of the memory interface that shows how support for atomic instructions is implemented. Please note that each core has its own separate memory interface. As already discussed in the preliminaries Sect. 2.2.2, the A instruction set extension provides two types of instructions: (1) AMO and (2) LR/SC. In the following we provide more details on how to implement these instructions and for illustration we will refer to Fig. 3.7.

```
1  .globl _start
2  .globl core_main
3
4  # NOTE: each core will start here with execution
5  _start:
6  csrr a0, mhartid # return a core specific number 0 or 1
7  li t0, 0
8  beq a0, t0, core0
9  li t0, 1
10 beq a0, t0, core1
11 # initialize stack for core 0 and core 1
12 core0:
13 la sp, stack0_end # code executed only by core 0
14 j end
15 core1:
16 la sp, stack1_end # code executed only by core 1
17 end:
18
19 # function argument stored in register a0 (according to RISC-V calling convention)
20 jal core_main
21
22 # wait until all two cores have finished
23 la t0, exit_counter
24 li t1, 1
25 li t2, 1
26 amoadd.w a0, t1, 0(t0) # get current counter value and increase existing value
27 # the first core reaching this point will spin
28 1:
29 blt a0, t2, 1b # jump one label backwards in case a0 < t2
30
31 # stop whole simulation with the *exit* system call
32 li a7, 93 # syscall exit has number 93
33 li a0, 0 # argument to exit
34 ecall # RISC-V system call
35
36 stack0_begin:
37 .zero 32768 # allocate 32768 zero-initialized bytes in memory
38 stack0_end:
39 stack1_begin:
40 .zero 32768
41 stack1_end:
42 exit_counter:
43 .word 0 # allocate 4 zero-initialized bytes in memory
```

Fig. 3.6 Bare-metal bootstrap code for a multi-core simulation with two cores

AMO Instructions To execute an AMO instruction, the core has to perform a load (Line 1) and store (Line 6) operation atomically without intervening memory access operations of other cores. We ensure the atomic execution property by locking the bus during the *amo* instruction. Therefore, a shared lock is acquired by the core's memory interface before the load operation (Line 2) and released again after the store (Line 52) operation. In case the bus is already locked, the lock operation will wait until the lock is released. Before performing a memory access operation, each core waits until it has obtained access rights (Line 35), i.e., the bus is not locked or is locked by the core itself. This locking scheme also supports DMI operations (Lines 39–45).

Peripherals that have write access to the bus (e.g., the DMA controller) are attached through a 1-to-1 TLM interconnect to the bus in order to ensure that they respect the bus locking. The peripheral interconnect transparently forwards all peripheral write transactions but waits in case the bus is locked by any core. Once the bus lock is released, all waiting (SystemC) processes are notified using a (SystemC) event.

LR/SC Instructions To execute an LR instruction, the core (memory interface) tracks a reservation on the load address and acquires the shared bus lock (Lines 12–13). The lock is kept acquired while *"forward-progress"* (see preliminaries Sect. 2.2.2) is maintained by the core. Essentially, the lock is released in case:

- More than 16 instructions are executed.
- A trap (exception) or interrupt is taken.
- A store is performed by the core holding the lock.

The SC instruction succeeds in case the lock is still acquired (Line 19) and a reservation on the store address exists (Line 20). In any case, the bus lock is released by the memory interface after executing the SC instruction (Line 24 or Line 52).

Similar to the AMO instructions, the shared bus lock ensures that other cores (and peripheral bus masters) do not interfere with the LR/SC instruction sequence execution and hence the LR/SC sequence eventually succeeds when maintaining *"forward-progress."*

3.1.5 VP Extension and Configuration

Our VP is designed as a configurable and in particular extensible platform. It is very easy to add additional components (i.e., peripherals/controllers including bus masters) and attach them to the bus system at a new address range or change the address mapping of the existing components. Furthermore, we provide a 32- and 64-bit core that can also be integrated into a multi-core platform. This allows for an easy (re-)configuration of the VP. By following the TLM 2.0 communication standard, transactions can be annotated with optional timing information to obtain a more accurate timing model of the executed software. Support for additional RISC-V ISA extensions (beyond IMAC) can be added inside the CPU core by extending the decode and execute functions accordingly. Our VP already provides a large feature set with large potential for additional extensions, which makes our VP suitable as foundation for different application areas.

In the following, we demonstrate the extensibility and configurability of our VP by three concrete examples: addition of a sensor peripheral, extension of a GDB-

```
1  int32_t atomic_load_word(uint64_t
       addr) {
2    bus_lock->lock(get_hart_id());
3    return load_word(addr);
4  }
5
6  void atomic_store_word(uint64_t addr,
       uint32_t value) {
7    assert (bus_lock->is_locked(
         get_hart_id() ));
8    store_word(addr, value);
9  }
10
11 int32_t atomic_load_reserved_word(
       uint64_t addr) {
12   bus_lock->lock(get_hart_id());
13   lr_addr = addr; // reservation for
         the load address on the whole
         memory
14   return load_word(addr);
15 }
16
17 bool atomic_store_conditional_word(
       uint64_t addr, uint32_t value) {
18   /* The lock is established by the
         LR instruction and the lock is
         kept while "forward-progress"
         is maintained. */
19   if (bus_lock->is_locked(
         get_hart_id() )) {
20     if (addr == lr_addr) {
21       store_word(addr, value);
22       return true; // SC succeeded
23     }
24     bus_lock->unlock();
25   }
26   return false; // SC failed
27 }
28 void store_word(uint64_t addr,
       uint32_t value) {
29   store_data(addr, value);
30 }
31
32 template <typename T>
33 inline void store_data(uint64_t addr,
       T value) {
34   // only proceed if the bus is not
         locked at all or is locked by
         this core
35   bus_lock->wait_for_access_rights(
         get_hart_id() );
36
37   // check if this access falls
         within any DMI range
38   bool done = false;
39   for (auto &e : dmi_ranges) {
40     if (e.contains(addr)) {
41       quantum_keeper.inc(
             e.dmi_access_delay );
42       *(e.get_mem_ptr_to_global_addr<T>(
             addr )) = value;
43       done = true;
44     }
45   }
46
47   // otherwise (no DMI), perform a
         normal transaction routed
         through the bus
48   if (!done)
49     do_transaction(
           tlm::TLM_WRITE_COMMAND, addr,
           (uint8_t *)&value, sizeof(T));
50
51   // do nothing in case the bus is
         not locked by this hart
52   bus_lock->unlock(get_hart_id());
53 }
54
55 // ... other load/store functions and
       load_data similar ...
```

Fig. 3.7 Core memory interface with atomic operation support

based SW debug functionality, and configuration matching the RISC-V HiFive1 board from SiFive.

3.1.5.1 Extending the VP with a Sensor Peripheral

This section presents the SystemC-based implementation of the VP sensor peripheral, which is used by the SW example presented in Sect. 3.1.2.1. It shows the principles in modeling peripherals and extending our VP as well as demonstrates the TLM communication and basic SystemC-based modeling and synchronization.

```
1  struct SimpleSensor : public
       sc_core::sc_module {
2    tlm_utils::simple_target_socket<
         SimpleSensor> tsock;
3
4    std::shared_ptr<InterruptController>
         IC;
5    uint32_t irq_number = 0;
6    sc_core::sc_event run_event;
7
8    // memory mapped data frame
9    std::array<uint8_t, 64> data_frame;
10
11   // memory mapped configuration
         registers
12   uint32_t scaler = 25;
13   uint32_t filter = 0;
14   std::unordered_map<uint64_t,
         uint32_t *> addr_to_reg;
15
16   enum {
17     SCALER_REG_ADDR = 0x80,
18     FILTER_REG_ADDR = 0x84,
19   };
20
21   SC_HAS_PROCESS(SimpleSensor);
22
23   SimpleSensor(sc_core::sc_module_name,
         uint32_t irq_number)
24     : irq_number(irq_number) {
25     tsock.register_b_transport(this,
           &SimpleSensor::transport);
26     SC_THREAD(run);
27
28     addr_to_reg = {
29       {SCALER_REG_ADDR, &scaler},
30       {FILTER_REG_ADDR, &filter},
31     };
32   }
33
34   void transport(
         tlm::tlm_generic_payload
         &trans, sc_core::sc_time
         &delay) {
35     // implementation available in
           Fig. 3.9
36   }
37
38   void run() {
39     while (true) {
40       run_event.notify(sc_core::sc_time(
             scaler, sc_core::SC_MS));
41       sc_core::wait(run_event); // 40
             times per second by default
42
43       // fill with random data
44       for (auto &n : data_frame) {
45         if (filter == 1) {
46           n = rand() % 10 + 48;
47         } else if (filter == 2) {
48           n = rand() % 26 + 65;
49         } else {
50           // fallback for all other
                 filter values: random
                 printable
51           n = rand() % 92 + 32;
52         }
53       }
54
55       IC->trigger_interrupt(irq_number);
56     }
57   }
58 };
```

Fig. 3.8 SystemC-based configurable sensor model that is periodically filled with random data—demonstrates the basic principles in modeling peripherals

The sensor is instantiated in the main function of the VP alongside other components and is attached to the TLM bus.

The sensor implementation is shown in Fig. 3.8. The sensor model has a data frame of 64 bytes that is periodically updated (overwritten with new data, Lines 44–53) and two 32-bit configuration registers *scaler* and *filter*. The update happens in the run thread (the run function is registered as SystemC thread inside the constructor in Line 26). Based on the scaler register value, this thread is periodically unblocked (Line 40) by calling the notify function on the internal SystemC synchronization event. Thus, *scaler* defines the speed at which new sensor data is generated. The *filter* register allows to select some kind of post-processing on the data. After every update, an interrupt is triggered, which will propagate through the interrupt controller to the CPU core up to the interrupt handler in the application

```
59  void transport(
        tlm::tlm_generic_payload &trans,
        sc_core::sc_time &delay) {
60    auto addr = trans.get_address();
61    auto cmd = trans.get_command();
62    auto len = trans.get_data_length();
63    auto ptr = trans.get_data_ptr();
64
65    if (addr >= 0 && addr <= 63) {
66      // access data frame
67      assert (cmd ==
            tlm::TLM_READ_COMMAND);
68      assert ((addr + len) <=
            data_frame.size());
69
70      // return last generated random
            data at requested address
71      memcpy(ptr, &data_frame[addr],
            len);
72    } else {
73      assert (len == 4); // NOTE: only
            allow to read/write whole
            register
74
75      auto it = addr_to_reg.find(addr);
76      assert (it != addr_to_reg.end());
            // access to non-mapped
            address
77
78      // trigger pre read/write actions
79      if ((cmd ==
            tlm::TLM_WRITE_COMMAND) &&
            (addr == SCALER_REG_ADDR)) {
80        uint32_t value =
            *((uint32_t*)ptr);
81        if (value < 1 || value > 100)
82          return; // ignore invalid values
83      }
84
85      // actual read/write
86      if (cmd == tlm::TLM_READ_COMMAND)
            {
87        *((uint32_t *)ptr) = *it->second;
88      } else if (cmd ==
            tlm::TLM_WRITE_COMMAND) {
89        *it->second = *((uint32_t *)ptr);
90      } else {
91        assert (false && "unsupported
            tlm command for sensor
            access");
92      }
93
94      // trigger post read/write actions
95      if ((cmd ==
            tlm::TLM_WRITE_COMMAND) &&
            (addr == SCALER_REG_ADDR)) {
96        run_event.cancel();
97        run_event.notify(sc_core::sc_time(
            scaler, sc_core::SC_MS));
98      }
99    }
100 }
```

Fig. 3.9 The *transport* function for the example in Fig. 3.8

SW. Therefore, the sensor has a reference to the interrupt controller (IC, Line 4) and an interrupt number provided during initialization (Lines 23 and 24).

Access to the data frame and configuration registers is provided through TLM transactions. These transactions are routed by the bus to the transport function (Line 34). The routing happens as follows: (1) The sensor has a TLM target socket field, which is bound in the main function (i.e., VP simulation entry point) to an initiator socket of the TLM bus. (2) The transport function is bound as destination for the target socket in the constructor (Line 25).

Based on the address and operation mode, as stored in the generic payload (Lines 60–61), the action is selected. It will either read (part of) the data frame (Line 71) or read/write one of the configuration registers (Lines 86–92). In case of a register access, a pre-read/write validation and post-read/write action can be defined as necessary. In this example, the sensor will ignore invalid scaler values (Lines 79–83) and reset the data generation thread on a scaler update (Lines 95–98). Please note that the transaction object (generic payload) is passed by reference and provides a pointer to the data, thus a write access is propagated back to the initiator of the transaction. Optionally, an additional delay can be added to the sc_time delay parameter (also passed by reference) for a more accurate timing model.

3.1.5.2 SW Debugging Support Extension

In this section we describe our VP extension to provide SW debug capabilities in combination with (for example) the Eclipse IDE by implementing the GDB *Remote Serial Protocol* (RSP) interface. Debugging for example enables to step through the SW line by line, set (conditional) breakpoints, obtain and even modify variable values, and also display the RISC-V disassembly (with the ability to step through the disassembly). Debugging can also be extremely helpful on the VP level to investigate errors, due to the deterministic and reproducible SW execution on the VP.

Using the GDB RSP interface, our VP acts as server and the GDB[4] as client. They communicate through a TCP connection and send text-based messages. A message is either a packet or a notification (a simple single char "+") that a packet has been successfully processed. Each packet starts with a "$" char and ends with a "#" char followed by a two-digit hex checksum (the sum over the content chars modulo 256). For example, the packet `$m111c4,4#f7` has the content m111c4,4 and checksum f7. The m command denotes a memory read, in this case read 0×4 bytes starting from address 0x111c4. Our server might then for example return `+$05000000#85`, i.e., acknowledge the packet (+) and return the value 5 (two chars per byte with little endian byte order). To handle the packet processing and TCP communication, we added a gdb-stub component to our VP. The whole debugging extension is only about 500 additional lines of C++ code most of them to implement the gdb-stub. On the VP side, only the CPU core has been modified to lift the SystemC thread into the gdb-stub, to allow the CPU to interrupt and exit the execution loop in case of a breakpoint and thus effectively transfer execution control to the gdb-stub.

Debugging works as follows: Start our VP in *debug-mode* (command line argument), this will transfer control to the gdb-stub implementing the RSP interface, waiting for a connection from the GDB debugger. In another terminal, start the GDB debugger. Load the same executable ELF file into the GDB (command "file main-elf") as in our VP. Connect to the TCP server of the VP (command "target remote :5005," i.e., to connect to local host using port 5005). Now the GDB debugger can be used as usual to set breakpoints, continue, and step through the execution. It is also possible to directly use a visual debugging interface, e.g., *ddd* or *gdb-dashboard* or even the *Eclipse* IDE.

Please note that the ELF file contains information about the addresses and sizes of the various variables in memory. Thus, a print(x) command with an int variable x

---

[4]In particular the freely available RISC-V port of the GDB, which knows about the available RISC-V register set, the CSRs, and can provide a disassembly of the RISC-V instruction set.

is already translated into a memory read command (e.g., m11080,4). Therefore, on the server side, i.e., our VP, an extensive parsing of ELF files is not necessary to add comprehensive debugging support. In total, we have only implemented 24 different commands of which nine can simply return an empty packet and a few more some predefined answer. Relevant packets are for example: read a register (p), read all registers (g), read memory range (m), set/remove breakpoint (Z0/z0), step (s), and continue (c).

3.1.5.3 HiFive1 Board Configuration

As an example, we provide the configuration matching the HiFive1 board from SiFive [108]. The HiFive1 is based around a SoC that integrates a single RISC-V RV32IMAC core with several peripherals and memories. Interrupts are processed by the CLINT and PLIC peripherals following the RISC-V ISA specification. The PLIC supports 53 interrupt sources with 7 priority levels. We configured our PLIC peripheral to match this specification. A (SPI) non-volatile flash memory is provided to store the application code and a small writable memory (DTIM) to hold the application data. The application data is initialized during the boot process by copying relevant data (e.g., initialized global variables) from the flash memory into the DTIM. This initialization code is embedded in the application binary that is placed in the flash memory. GPIOs and an UART are provided for communicating with the environment. We redirect each write access to the UART to stdout. We used our register modeling layer to create these additional peripheral models in SystemC. For the GPIO peripheral, we provide a (re-usable) interface to access an environment model. We provide more details on this interface in the following section.

GPIO Environment Interaction Interface GPIOs are used to connect the embedded system to the outside world. Each GPIO pin can be configured to serve as output or input connection. Input pins can trigger an interrupt when being written.

We integrate a GPIO-(TCP) server to provide access to the GPIO pins in order to attach an environment model. The server runs in a separate system thread and hence needs to be synchronized with the SystemC kernel. Therefore, we use the SystemC *async_request_update* function to register an update function to be executed on the next update cycle of the SystemC thread. The update function triggers a normal SystemC event notification to wake up the SystemC GPIO processing thread.

Figure 3.10 shows an example virtual environment implemented in C++ using the Qt Framework. It shows the HiFive1 board with a seven-segment display and a button attached to the GPIO pins. The display shows the current counter value, and the button is used to control the count mode (switch between increment and decrement). We have also executed the same RISC-V ELF binary on the real HiFive1 board using the same setup as shown in the virtual environment.

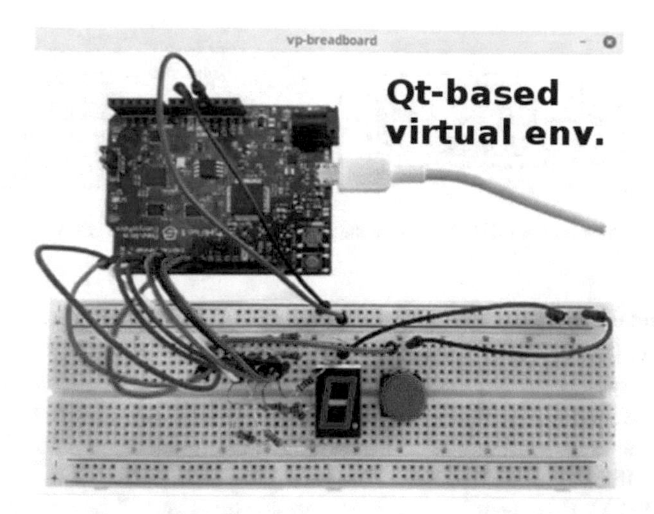

Fig. 3.10 Qt-based virtual environment, showing the HiFive1 board with a seven segment display (output) and a button (input), attached to the VP simulation through a TCP connection

3.1.6 VP Evaluation

In this section we first describe how we have tested our VP to evaluate and ensure the VP quality, then we present results of a performance evaluation.

3.1.6.1 Testing

Testing is very important to ensure that the VP is working correctly. In the following we describe how we have tested our VP. In particular, we used

1. The official RISC-V ISA tests [181], in particular the RV32/64IMAC tests.
2. Use the RISC-V Torture test-case generator [182], also targeting the RV32/64IMAC instruction set.
3. Employ state-of-the-art coverage-guided fuzzing (CGF) techniques for test-case generation. For more information on this approach, please refer to [102].
4. Several example applications, ranging from bare-metal applications to examples using the FreeRTOS [60] and Zephyr [224] operating systems. We observe if the example applications behave as expected.

Please note that the RISC-V ISA tests are directed tests that are handwritten and already encode the expected result inside of the test. Hence, no reference simulation is required for the ISA tests.

Figure 3.11 shows an overview of the Torture and CGF approach. They both work by generating a set of tests, i.e., each testcase is a RISC-V ELF binary, which is then executed one after another on our VP and (one or multiple) reference simulators. We

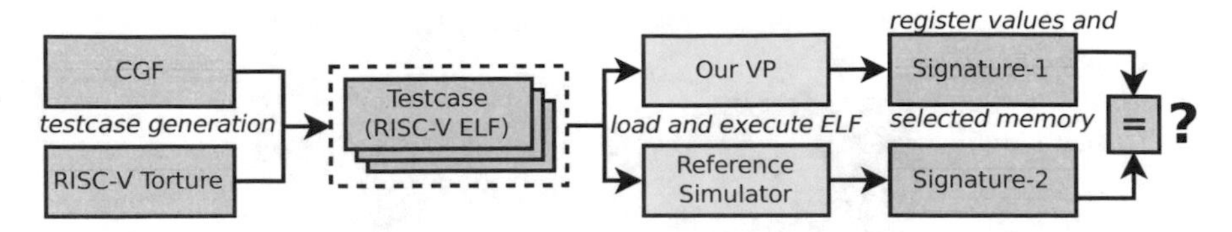

Fig. 3.11 Overview on the RISC-V Torture and CGF approach for VP testing

use the official RISC-V reference simulator Spike [197] in this evaluation. We have extended our VP to dump the execution result, called *signature*, as supported by Spike. The *signature* contains the register values and selected memory contents. After each execution, the dumped signatures are compared for differences. Hence, the simulator is considered as a black box and no intrusive modifications are required for this testing.

The RISC-V ISA tests as well as the Torture and Fuzzer test generation approaches primarily focus on testing the core. The example applications however also tend to analyze larger portion of the whole VP platform. They further integrate the CLINT and PLIC interrupt controller as well as selected peripherals that are required for the particular application. In the following we provide more details on some selected applications that demonstrate the applicability of our VP for real-world embedded applications.

Zephyr Examples The Zephyr OS is designed around a small-footprint kernel primarily targeting (resource constrained) embedded systems. Zephyr supports multiple architectures, including RISC-V. For the RISC-V architecture, support for the HiFive1 board is available. We used the HiFive1 board configuration of our VP to run several Zephyr example applications that extensively use core components of the kernel, including threads, timers, semaphores, and message queues. In addition, we run examples that perform aes128 and sha256 encryption/decryption using the TinyCrypt library.

FreeRTOS Examples Similar to Zephyr, FreeRTOS is also designed around a kernel component, targets embedded systems, and provides support for the RISC-V architecture. We also build several example applications that create multiple threads with different priorities, use queues for data passing, and integrate interrupts. In addition, we created two applications that use the FAT and UDP library extensions of FreeRTOS. The first application formats an SD card by creating a new MBR and writing an FAT partition. The second application sends/receives a set of UDP packets using our ethernet peripheral to/from a remote computer.

3.1.6.2 Performance Evaluation

In our paper [96] we have already demonstrated: (1) that our VP provides more than 1000x faster simulation performance compared to an RTL implementation, and (2) the effectiveness of our presented VP simulation performance optimization techniques (between 6.1x and 7.8x improvement on the considered benchmark set). This book presents an updated performance evaluation of our VP on a more recent and faster simulation host. We use a set of different applications to demonstrate the execution performance, measured in MIPS (Million Instructions Per Second), of our VP. Furthermore, we provide a performance comparison to other RISC-V simulators, in particular: FORVIS, SAIL (the C simulator back-end in particular), gem5, SPIKE, and QEMU

Experiment Setup and VP Results All experiments are performed on a Fedora 29 Linux system with an Intel Xeon Gold 5122 processor with 3.6 GHz. The memory limit is set to 32 GB and the timeout to 4 h (i.e., 14,400 s). Our VP is compiled with GCC 8.2.1 with -O3 optimization.

Table 3.1 shows the results. The first six columns show the benchmark name (column: *Benchmark*), number of executed instructions (column: *#instr*), *Lines of Code* (LoC) in C (column: *C*) and in assembly (column: *ASM*), the aforementioned MIPS (column: *MIPS*) performance metric, and the simulation time (column: *time*) for our VP. The remaining five columns show the simulation time for the other RISC-V simulators. All simulation times are reported in seconds. M.O. denotes a memory out and N.S. that the benchmark is not supported by the simulator. We consider benchmarks from different application areas:

- *dhrystone* is a synthetic computing benchmark designed to measure the general (integer) execution performance of a CPU. We execute 10,000,000 iterations of the algorithm.
- *qsort* is an efficient sorting algorithm. We use a standard implementation to sort an array with 50,000,000 elements.
- *fibonacci* is a small program implemented in assembler that performs 1,000,000,000 iterations and ignores any integer overflow.
- *zephyr-crypto* uses two threads that communicate through a (single-element) message queue. The first thread encrypts an ~1 MB data set, using the aes128 algorithm of the TinyCrypt library, the second thread decrypts it again.

Table 3.1 Experiment results—all execution times reported in seconds and number of executed instructions (#instr) reported in Billions (B)

		LoC		VP		Other simulators (Time)				
Benchmark	#instr	C	ASM	MIPS	Time	FORVIS	SAIL	gem5	SPIKE	QEMU
dhrystone	4.06 B	362	2 212	42.7	94.86	M.O.	10 986.24	1 170.08	13.90	3.53
qsort	2.93 B	146	2 279	42.8	68.71	M.O.	T.O.	985.99	11.01	3.04
fibonacci	5.99 B	/	17	53.9	111.15	13 954.57	12 830.05	1 447.37	17.19	1.94
zephyr-crypto	2.52 B	86	5 407	46.6	54.11	N.S.	N.S.	N.S.	N.S.	6.50
mc-ivadd	2.39 B	26	1 798	44.4	53.76	N.S.	N.S.	N.S.	48.11	N.S.

MIPS = Millions Instructions Per Second. LoC = Lines of Code in C and assembly (ASM). M.O. = Memory Out (32GB limit). T.O. = Time Out (4h = 14,400 s limit). N.S. = Not Supported

- *mc-ivadd* is a multi-core benchmark that performs a vector addition and stores the result in a new vector. Each core operates on a different part of the vector. The vector size is set to 4,194,304 and the algorithm performs 30 iterations.

We report results for using our VP with an RV32 core (four RV32 cores for the multi-core benchmark). However, similar results are obtained when using an RV64 core(s).

It can be observed that our VP provides a high simulation performance between 42 and 53 MIPS with an average of 46 MIPS on our benchmark set, which demonstrates the applicability of our VP to real-world embedded applications. The pure assembler program (fibonacci) achieves the highest simulation performance, since it does not require to perform memory access operations (besides instruction fetching). It can also be observed that the additional synchronization overhead (in the SystemC simulation) to perform a multi-core simulation has no significant performance impact[5] Please note that, for the multi-core simulation benchmark, we report the total MIPS for all four cores.

Comparison with Other Simulators Compared to the other RISC-V simulators (right side of Table 3.1), our VP shows very reasonable performance results and is located in the front midfield. As expected, our VP is not as fast as the high-speed simulators SPIKE and in particular QEMU. The reason is the performance overhead of the SystemC simulation kernel and the more detailed simulation of our VP.[6] However, our VP is much faster than FORVIS and SAIL as well as gem5, which arguably is the closest to our VP in terms of the intended use cases. Please note that the *zephyr-crypto* and *mc-ivadd* benchmarks are not supported (N.S.) on some simulators due to missing support for the Zephyr operating system and the multi-core RISC-V test environment, respectively. In the following we discuss the results in more detail.

Compared to QEMU, the performance overhead is most strongly pronounced on the *fibonacci* benchmark. The reason is that the benchmark iterates a single basic block and only performs very simple operations (mostly additions) without memory accesses. This has two implications: This single basic block is pre-compiled once into native code using DBT and then reused for all subsequent iterations (hence QEMU executes at near native performance). Furthermore, since only simple operations are used, the simulation overhead induced by SystemC has a comparatively strong influence on the overall performance.

On more complex benchmarks such as *qsort* and *zephyr-crypto*, the overhead to QEMU is less strongly pronounced. These benchmarks have a much more complex control flow and hence require to compile several basic blocks and also use more

[5]Please note that the SystemC kernel is not using a multi-threaded simulation environment but executes one process (i.e., one core) at a time and switches between the processes.

[6]This includes more accurate timing by leveraging SystemC, instruction accurate interrupt handling, and the ability to integrate TLM-2.0 memory transactions. Furthermore, compared to QEMU we currently do not integrate DBT (Dynamic Binary Translation).

complex instructions that take longer time for (native) execution (compared to a simple addition). Furthermore, *zephyr-crypto* performs several simulated context switches between the Zephyr OS and the worker threads, which has additional impact on the simulation performance of QEMU (since it causes an indirect and non-regular transfer between the basic blocks).

Compared to SPIKE, the performance overhead is mostly uniformly distributed (since SPIKE does not use DBT either) on the single core benchmarks (with SPIKE being around 6.2x–6.8x faster than our VP). The performance on the multi-core benchmark *mc-ivadd* however is very similar for both SPIKE and VP. Apparently, SPIKE is very strongly optimized for simulation of single core systems and hence the newly introduced synchronization and context switch overhead is very significant.

Compared to gem5, our VP is significantly faster (between 12.3x to 14.4x). The primary use case of gem5 is also not a pure functional simulation (which is the goal of SPIKE and QEMU) but rather an architectural exploration and analysis. Thus, gem5 provides more detailed processor and memory models with complex interfaces, which in principle can also be extended for accurate extra-functional properties. Furthermore, gem5 is a large (and aims to be a rather generic) platform, which supports different architectures besides RISC-V, which causes additional performance overhead.

Compared to FORVIS and SAIL, our VP is consistently much faster (up to two orders of magnitude). The reason is that these simulators have been designed with a different use case in mind (establishing an executable formal representation of the RISC-V ISA) and hence very fast simulation performance is only a secondary goal. Furthermore, FORVIS runs into memory outs (M.O.), which may indicate a memory leak in the implementation. SAIL has a time out (T.O.) on the *qsort* benchmark.[7]

In summary, the simulation performance very strongly depends on the simulation technique, which in turn depends on the primary use case of the simulator. The goal of our VP is to introduce an (open-source) implementation that leverages SystemC TLM-2.0 into the RISC-V ecosystem to lay the foundation for advanced SystemC-based system-level use cases. Overall, our VP provides a high simulation performance with an average of 46 MIPS on our benchmark set.

3.1.7 Discussion and Future Work

Our RISC-V based VP is implemented in SystemC TLM 2.0 and already provides a significant set of features, which makes our VP suitable as foundation for various application areas, including early SW development and analysis of interactions at

[7]A 10x smaller version of *qsort* was completed successfully within 700.22 s on SAIL, and hence this time out may also indicate a memory-related problem in SAIL (since *qsort* requires a significant amount of memory for the large array to be sorted).

the HW/SW interface of RISC-V-based systems. Recently, we extended our VP to provide support for RISC-V floating-point extensions and integrated an MMU (Memory Management Unit) to support virtual memory and memory protection. This, extensions enabled our VP to boot the Linux OS. Nonetheless, our VP can still be further improved. In the following we sketch different directions that we consider for future work.

One direction is the extension of our VP with new components and integration of additional RISC-V ISA extensions. In particular, dedicated support for custom instruction set extensions is very important. We plan to look into appropriate specification mechanisms and develop suitable interface to facilitate specification, generation, and integration of custom RISC-V extensions.

Performance optimizations are also very important, in particular to run a whole SW stack including the Linux OS. Two techniques seem very promising: (1) Integration of DBT/JIT (Dynamic Binary Translation/Just-In-Time) techniques to avoid the costly interpreter loop whenever possible, e.g., translate and cache RISC-V basic blocks.[8] (2) Use a real thread for each core in a multi-core simulation. This requires dedicated techniques to synchronize with the SystemC simulation.

While we have already performed an extensive testing of our VP, in particular the core, we plan to consider additional verification techniques and also put a stronger emphasis on verification of peripherals and other IP components. In particular, we plan to consider formal verification techniques for SystemC, e.g., [38, 100, 212], and also investigate (UVM-based) constrained random techniques for test-case generation [222].

Finally, we want to provide support for SW debugging of multi-core platforms by extending our implementation of the GDB RSP interface accordingly.

The next section, presents an efficient core timing model, which we have integrated into the RISC-V VP core, to enable fast and accurate performance evaluation for RISC-V-based systems.

3.2 Fast and Accurate Performance Evaluation for RISC-V

As mentioned already, there exist a number of RISC-V simulators such as the reference simulator Spike [197], RISCV-QEMU [185], RV8 [186], DBT-RISE [42], or Renode [176]. These simulators differ in their implementation techniques and their intended use case, which range from mainly pure CPU simulation (RV8, Spike) to full-system simulation (RISCV-QEMU, DBT-RISE) and even support for multi-node networks of embedded systems (Renode). However, they are mainly designed to simulate as fast as possible to enable efficient functional validation

[8]This (dynamic) translation from RISC-V instructions to the simulation host instruction set should also preserve (for example) timing information of the SystemC simulation to avoid losing accuracy in the simulation timing model.

of large and complex systems and predominantly employ DBT (*Dynamic Binary Translation*) techniques, i.e., translate RISC-V instructions on the fly to *x86_64*. This is however a trade-off as accurately modeling timing information becomes much more challenging and thus these simulators do not offer a cycle-accurate performance evaluation.

A full-system simulator that can provide accurate performance evaluation results and recently got RISC-V support is *gem5* [18]. However, *gem5* integrates more detailed functional models (e.g., of the CPU pipeline) instead of abstract timing models that are sufficient for a pure performance evaluation and hence the simulation performance is significantly reduced.

Commercial VP tools, such as Synopsys Virtualizer or Mentor Vista, might also support RISC-V in combination with fast and accurate timing models but their implementation is proprietary.

Many recent state-of-the-art performance evaluation techniques focus on *Source-Level Timing Simulation* (SLTS) to enable a high-speed performance evaluation [26, 145, 165, 198, 215, 216]. SLTS works by instrumenting the application source code with timing information that are typically obtained by a static analysis, and then host compiles the instrumented source code and natively executes the resulting binary without any emulation layer. However, due to the source-level abstraction it is very challenging to provide accurate models for peripherals and consider complex HW/SW interactions such as interrupts accurately.

Other recent approaches leverage DBT-based techniques. The papers [31, 33] discuss a combination of QEMU and the SystemC kernel. However, these methods either only provide rough performance estimates or incur significant synchronization overhead between QEMU and SystemC and lose the determinism of a SystemC simulation. Another DBT-based approach is [24], which integrates a timing model with a DBT-based execution engine to obtain near cycle accurate results. The work by Thach et al. [206] is conceptually similar but uses QEMU as execution back-end. However, neither of these approaches are VP-based or target the RISC-V ISA.

In the following we present an efficient core timing model and its integration into the RISC-V VP core, to enable fast and accurate performance evaluation for RISC-V-based systems.

3.2.1 Background: HiFive1 Board

The HiFive1 board from SiFive has a 32-bit RISC-V core that integrates a 16-KiB 2-way set-associative instruction cache with a branch prediction unit and a single-issue, in-order pipeline with five stages. The pipeline can achieve a peak execution rate of one instruction per cycle. However, multiplication and division as well as load and store instructions require additional cycles to provide the result and hence can stall the pipeline. The branch prediction unit has a 40-entry *Branch Target Buffer* (BTB), a 128-entry *Branch History Table* (BHT), and 2-entry *Return Address Stack* (RAS). We will describe these branch prediction components as part of our branch

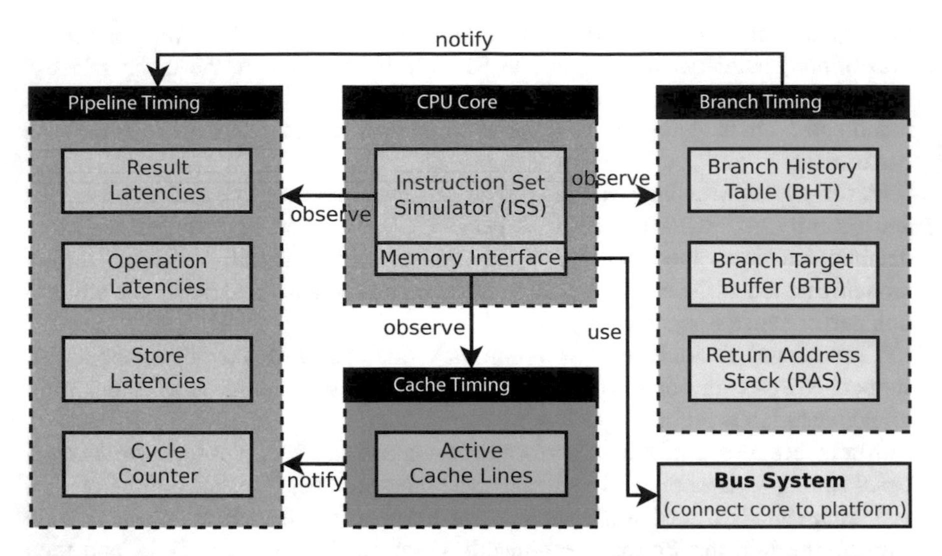

Fig. 3.12 Overview on our core timing model and the integration with the CPU core model

prediction timing model in Sect. 3.2.2.3 in more detail. Correctly predicted branches result in no execution penalty, but mispredicted branches incur a three-cycle penalty. Instructions can be fetched with a one-cycle latency from the cache. A cache miss triggers a burst access to load the corresponding cache line (32-byte). Please note that the HiFive1 board has no designated DRAM but provides a small cache-like data memory that has access rates with fixed latencies (depending on the access size). However, nearby load/store instructions with overlapping addresses result in additional execution penalties. For more technical details, please refer to the official manual [195].

3.2.2 *Core Timing Model*

In this section we present our core timing model that enables a fast and accurate performance evaluation in combination with VPs. We start with an overview. Then, we present more details on our branch prediction (Sect. 3.2.2.3), pipeline (Sect. 3.2.2.2) and cache (Sect. 3.2.2.4) timing models.

3.2.2.1 Overview

Figure 3.12 shows an overview on our core timing model. Essentially, it consists of three parts: a pipeline timing model (left side Fig. 3.12), a branch (predictor) timing model (right side Fig. 3.12), and a cache timing model (bottom Fig. 3.12). These timing models allow to consider timing information that depends on pipelining, branch prediction, and caching effects, respectively. Please note that in this work we do not consider effects of out-of-order execution, since this is typically not so common for embedded systems.

All three timing models are attached to the CPU core model, which essentially consists of an ISS and a memory interface. The ISS fetches, decodes, and executes instructions one after another. The memory interface performs all memory access operations for the ISS. In particular it is responsible to perform the actual fetch of the next instruction as well as perform load and store operations. The memory interface typically wraps all operations into TLM bus transactions and forwards them to the bus system, which then routes the transactions to the actual target peripherals (hence connects the ISS with the rest of the VP platform).

We observe all executed instructions in the ISS in order to update our pipeline and branch prediction models accordingly. The branch predictor is only queried and updated on branch and jump instructions. The pipeline needs to be updated for instructions that do not finish within a single execution cycle, as they can introduce latencies in case the next instructions depend on the result of a previous instruction before it is available. In this case the pipeline needs to be *stalled*. Typical instruction that require multiple cycles to completion are multiplication/division as well as load/store instructions and instructions that access special registers. We also observe the memory interface in order to update our cache timing model.

Both the branch prediction timing model and the cache timing model have access to the pipeline timing model. In case of a branch misprediction or cache miss, the pipeline is effectively *stalled*. This will update any pending result latencies in the pipeline and increase the cycle counter accordingly.

Please note that it is not necessary to build timing models that are fully functional but only sufficient to keep track of timing-related information. For example, a timing model for the cache needs only to decide if a cache hit or miss occurs, therefore it is not necessary to keep track of the actual cache contents but only keep track of the currently cached address ranges. Another example is the timing model for the pipeline, which needs only enough information to decide if the pipeline will stall and for how many cycles. Therefore, it is typically not necessary to keep track of all the pipeline stages for each instruction, but only result dependencies between instructions and their computation latencies. By performing these (functional) abstractions in the timing models, the computational complexity is significantly reduced, and hence also the performance overhead minimized for keeping track of the timing information, while a very high accuracy is obtained. In the following we describe our three timing models in more detail.

3.2.2.2 Pipeline Timing Model

The pipeline timing model keeps track of timing relevant execution dependencies between instructions in order to decide when and for how many cycles to stall the pipeline. A stall of the pipeline corresponds to *advancing* the pipeline timing model and is implemented by simply increasing the (total execution) cycle counter and accordingly decreasing the (local latency) counters of any pending dependencies (for illustration we show an example later in this section). We consider three kinds of dependencies: *result*, *operation*, and *store*.

A result dependency denotes that an instruction N is accessing a register R that a previous instruction P writes to and P takes more than one cycle to compute the result, and hence N has to wait for R. In this case we add a result latency for R to our pipeline model. In case a register is read/written (during instruction execution), which has a result latency X, the pipeline model is stalled for X cycles. For better illustration, Fig. 3.13 shows an example. It shows four instructions, two additions (*add*) and two multiplications (*mul*) where *mul* has a 5-cycle result latency and *add* has no result latency (i.e., the result of *add* is directly available for the next instruction in the pipeline). Figure 3.13 (right side) reports the current cycle counter and result latencies after the instruction has finished. The cycle counter starts with zero, and the result latencies are empty. Both *mul* instructions add a 5-cycle result latency on their destination register. After each instruction, the pipeline is advanced one clock cycle, updating the cycle counter and pending result latencies accordingly. The fourth instruction additionally stalls the pipeline model for 4 cycles due to the pending 4-cycle result latency on register a1. Pending operation and store latencies are handled similarly to result latencies.

An operation dependency denotes that an instruction requires a specific resource and hence reserves it for a specific time T. Other instructions that require the same resource need to wait until the resource is released, i.e., T cycles pass. For example, the RISC-V HiFive1 board has only a single multiplication unit, which has a 5-cycle result latency.

A store dependency denotes that an instruction has performed a memory write access, which is scheduled to be committed after X cycles. Load instructions, which access an overlapping memory address range before X cycles passed, will be delayed for a specific configurable time by stalling the pipeline.

Please note, many instructions—like addition, shift, bit operations, etc.—can be effectively executed in a single cycle on embedded systems with a pipeline. For these common instructions, no updates are necessary in the pipeline timing model, and hence they effectively incur no performance overhead (besides incrementing

Operation	Instr.	Reg. Operands			Cycle Counter	Result Latencies
		RD	RS1	RS2		
1: a0=a1+a2	add	a0	a1	a2	1	[]
2: a1=a0*a1	mul	a1	a0	a1	2	[a1 -> 5]
3: a0=a0+a2	add	a0	a0	a2	3	[a1 -> 4]
4: a2=a1*a2	mul	a2	a1 #	a2	8	[a2 -> 5]

stall for 4 cycles due to the pending result latency on a1

Fig. 3.13 Example to illustrate the pipeline timing model

```
1  struct BranchPredictionTiming {
2    static constexpr uint
         BranchMispredictionPenalty = 3;
3    std::shared_ptr<PipelineTiming>
         pipeline;
4    DefaultRAS RAS; // Return Address
         Stack
5    DefaultBTB BTB; // Branch Target
         Buffer
6    DefaultBHT BHT; // Branch History
         Table
7
8    void jump(uint instr_pc, uint
         target_pc) {
9      if (BTB.find(instr_pc) !=
           target_pc) { // wrong BTB
           prediction
10       BTB.update(instr_pc, target_pc);
11       pipeline->stall(
             BranchMispredictionPenalty);
12     }
13   }
14
15   void call(uint instr_pc, uint
         target_pc, uint link_pc) {
16     jump(instr_pc, target_pc);
17     RAS.push(link);
18   }
19
20   void ret(uint instr_pc, uint
         target_pc) {
21     if (RAS.empty() || (target_pc !=
           RAS.pop()))
22       pipeline->stall(
             BranchMispredictionPenalty);
23   }
24
25   void branch(uint instr_pc, uint
         target_pc, bool branch_taken) {
26     Entry &entry =
           BHT.get_entry(instr_pc);
27     if (entry.predict_branch_taken()){
28       // predict that branch will be
             taken
29       if (branch_taken) {
30         jump(instr_pc, target_pc); //
               correct BHT prediction
31       } else { // wrong BHT prediction
32         pipeline->stall(
               BranchMispredictionPenalty);
33       }
34     } else {
35       // predict that branch will not
             be taken
36       if (branch_taken) { // wrong BHT
             prediction
37         BTB.update(instr_pc,
               target_pc);
38         pipeline->stall(
               BranchMispredictionPenalty);
39       }
40     }
41     bht_entry.update_history(
           branch_taken);
42   }
43 };
```

Fig. 3.14 Branch predictor timing model that provides four interface functions (*jump*, *call*, *ret*, and *branch*) to update the timing model after a jump, function call/return, and branch instruction, respectively

the cycle counter) on our timing model in particular if no active dependencies are pending (who would need to be updated).

3.2.2.3 Branch Prediction Timing Model

Figure 3.14 shows our timing model for our branch predictor. The model is updated on branch (Line 25) and jump (Line 8) instructions. For jump instructions, we also distinguish the special cases of a (function) call (Line 15) and return (Line 20).

In case of a branch, the model has to predict if the branch will be taken and in case of a taken branch the address of the target instruction. For a jump (and its variants: call and return), it has to predict the target of the jump. In case of a misprediction, an execution time penalty is added by *stalling* the pipeline timing model (which mimics the timing effects of a pipeline flush due to the misprediction).

To perform the predictions, the branch predictor timing model manages three sub-models: BTB (*Branch Target Buffer*), BHT (*Branch History Table*), and RAS (*Return Address Stack*). For the following description, as well as in Fig. 3.14, we use the following naming convention: *instr_pc* refers to the address of the branch/jump instruction, *target_pc* refers to the address of the branch/jump target, and *link_pc* is the address of the instruction following the branch/jump instruction in the code (due to potentially varying instruction lengths, *link_pc* typically cannot be inferred based on *instr_pc*).

The BTB is simply a fixed size mapping from *instr_pc* to (the predicted) *target_pc*. The size and replacement strategy is configurable. By default the lower bits of *instr_pc* are used to index the mapping. The BTB is queried to predict the target of branch and jump instructions (Lines 9 and 30) and updated in case of mispredictions (Lines 10 and 37).

The RAS manages a fixed size stack of return addresses for the last function calls. RAS is updated on every *call* instruction (Line 17) and checked on every *return* instruction (Line 21). In case of a misprediction, an execution latency can be directly incurred (Line 22) or an alternative would be to query the BTB (i.e., call the *jump* function in Line 22 instead). RAS drops old entries in case a new entry is pushed on a full stack (Line 17).

The BHT keeps track of the recent execution history for branch instructions, i.e., whether the branch has been taken or not, to predict the outcome of subsequent branches. The BHT is queried (Line 27) and updated (Line 41) on every branch instruction. In case of a correct BHT prediction, the BTB is queried (Line 30 by calling *jump*) to predict the actual branch target. In case of a BHT misprediction, an execution cycle penalty is incurred by stalling the pipeline timing model (Lines 32 and 38). Our BHT implementation has a fixed size mapping indexed by *instr_pc* (i.e., using the lower bits of the address of the branch instruction). Each mapping entry stores the recent (execution) history and a set of (prediction) counters for every execution history variant. The size of the mapping as well as the length of the execution history and counters is configurable.

As an example, consider a BHT configuration with two history bits and two counter bits for each mapping entry. Two history bits allow to distinguish four different cases (by interpreting a 1 as branch Taken and a 0 as Not taken). Thus, four two-bit counters are provided (one for each possible history case). Each two-bit counter can store four possible values: 0 (strongly Not taken), 1 (Not taken), 2 (Taken), and 3 (strongly Taken). Half of the values result in a branch Taken prediction (2,3) and the other half a branch Not taken prediction (0,1). Thus, Line 27 essentially checks if *counters[history] >= 2* and the update in Line 41 basically increments or decrements the *counters[history]* and shifts the history, depending on the actual observed branch direction, i.e., the *branch_taken* variable. Figure 3.15 shows a concrete example for further illustration. It shows a simple C/C++ loop (left top) with corresponding (RISC-V) assembler code (left bottom) and a table that shows relevant data for the loop branch instruction (right side). It shows the

```
i = 1;
while (i < 10)
  ++i;
```

```
li a0,1
li a1,10
loop:
bge a0,a1,end
addi a0,a0,1
j loop
end:
```

loop-branch BHT entry		Init	Loop Iteration				
			1	2	3	...	10
Prediction		/	T	N	N	N	N
Observation		/	N	N	N	N	T
Counters	NN	1	1	1	0	0	1
	NT	1	1	0	0	0	0
	TN	3	3	3	3	3	3
	TT	3	2	2	2	2	2
History		TT	NT	NN	NN	NN	TN

Fig. 3.15 Example to illustrate the branch timing model

current history and counter values for the BHT entry of the loop branch after each loop iteration (the initial state is chosen arbitrarily in this example) as well as the predicted and really observed branch direction. Based on the BHT entry in iteration N, a prediction is made for iteration N+1. For example, after iteration 1 the history is NT, and hence the NT counter decides the prediction N due to the counter value 1. Based on the observation N in iteration 2, the counter NT is updated accordingly (decremented) and the N is shifted into the history to obtain the BHT entry state after iteration 2.

Besides several configuration options, like the execution history and prediction counter lengths, this branch prediction timing model can also be easily adapted by, e.g., simply removing the RAS in case the embedded system does not cache function call return addresses.

3.2.2.4 Cache Timing Model

The cache timing model is queried on memory access operations to cached regions. Typically, an embedded system provides at least an instruction cache, and hence the timing model will be queried for every instruction fetch. Thus, the timing model should be very efficient in deciding a cache hit or miss to reduce the performance overhead. In case of a cache miss, an execution penalty is added by *stalling* the pipeline timing model (which mimics the timing effects of waiting for a cache line fetch).

We provide a model matching a two-way associative cache with LRU replacement strategy, which is a common choice for embedded systems. However, our cache timing model can be easily generalized to an N-way associative cache with a different replacement strategy. The cache table consists of cache lines, which in turn consist of two entries (for a two-way associative cache). We only keep track of the memory address that the entries represent and not the data, since the actual data is not relevant for our timing model. The number of lines and their size in bytes is configurable. Each memory access address is translated to an index into the cache table by taking the lower bits of the address. Finally, we compare the memory access address with the address of both cache line entries to decide if the memory access is a cache hit or miss. In case of a miss (none of the entries does match), one of the two entries is updated (i.e., associate the entry with a new memory address range, which matches the memory access address) based on the replacement strategy.

3.2.3 Experiments

We have implemented our proposed core timing model and integrated it into the open-source RISC-V VP (see Sect. 3.1). Furthermore, we configured our core timing model to match the RISC-V HiFive1 board from SiFive (see Sect. 3.2.1). In this section we present experiments that evaluate the accuracy and simulation performance overhead of our core timing model. We obtain accuracy results by comparing against a real RISC-V HiFive1 board from SiFive, and we obtain the performance overhead results by comparing against the RISC-V VP without integrating our core timing model. All experiments are evaluated on a Linux machine with Ubuntu 16.04.1 and an Intel i5-7200U processor with 2.5 GHz (up to 3.1 GHz with Turbo Boost).

For evaluation, we use the *Embench* benchmark suite [50, 51].[9] It is a freely available standard benchmark suite specifically targeting embedded devices. The *Embench* benchmark suite is a collection of real programs instead of synthetic programs like *Dhrystone* or *Coremark* to ensure that more realistic workloads are considered. For example, it contains benchmarks that perform error checking, hashing, encryption and decryption, sorting, matrix multiplication and inversion, JPEG and QR-code as well as regex and state machine operations. The benchmarks have a varying degree of computational, branching, and memory access complexity [15]. Each benchmark starts with a *warm-up* phase that executes the benchmark body a few times to warm up the caches. Then, the real benchmark starts by executing the benchmark body again several more times. The number of iterations is configurable and depends on the CPU frequency. Furthermore, it varies between different benchmarks to ensure a mostly uniform runtime complexity across the benchmark suite. We set the CPU frequency constant to 320 MHz to match the HiFive1 board.

Table 3.2 shows the results. Starting from the left, the columns show the benchmark name (Column 1), the *Lines of Code* (LoC) in C (Column 2) and RISC-V Assembler (Column 3), and the number of executed RISC-V instructions (Column 4, measured by the RISC-V VP). Then, Table 3.2 shows the simulation time in seconds for running the benchmark on the RISC-V VP without (Column 5) and

[9]We omitted three of the nineteen benchmarks from the comparison due to problems in executing them on the real HiFive1 board.

Table 3.2 Experiment results—all simulation time results reported in seconds

	LoC			VP	VP + core timing model		HiFive1 board	Difference	
1: Benchmark	2: C	3: ASM	4: #Instrs.	5: Time	6: Time	7: #Cycles	8: #Cycles	9: #Cycles	10: Time
aha-mont64	306	4887	1,083,316,249	27.46	32.05	1,130,126,565	1,192,518,796	−5.2%	+16.7%
crc32	256	3890	1,093,865,270	30.02	33.51	1,283,897,183	1,284,037,753	−0.0%	+11.6%
edn	440	4655	1,009,188,407	30.81	34.12	1,772,167,527	1,847,419,217	−4.1%	+10.7%
matmult-int	330	4076	1,039,623,303	30.61	33.68	1,761,624,764	1,780,530,752	−1.1%	+10.0%
minver	343	6262	948,801,750	32.90	39.78	1,346,343,318	1,298,778,561	+3.7%	+20.9%
nbody	322	6427	995,233,372	32.25	39.18	1,356,400,421	1,364,440,490	−0.6%	+21.5%
nettle-aes	1173	5831	1,203,436,482	34.82	42.10	1,219,069,707	1,223,022,326	−0.3%	+20.9%
nettle-sha256	503	6252	1,078,331,659	36.65	43.57	1,101,319,048	1,106,544,095	−0.5%	+18.9%
picojpeg	2337	9482	957,883,248	30.23	36.09	1,208,694,004	1,158,481,492	+4.3%	+19.4%
qrduino	1091	7726	1,142,278,239	34.28	38.58	1,399,022,255	1,544,185,605	−9.4%	+12.5%
sglib-combined	2002	6501	906,977,088	29.28	35.69	1,190,579,040	1,209,895,484	−1.6%	+21.9%
slre	661	5518	1,110,788,540	35.71	43.69	1,443,734,072	1,272,198,737	+13.5%	+22.3%
st	271	6643	1,051,545,074	33.60	40.94	1,425,869,796	1,333,455,328	+6.9%	+21.8%
statemate	1456	5866	789,501,877	27.68	38.04	988,087,257	1,077,693,835	−8.3%	+37.4%
ud	250	4774	841,910,324	27.02	30.55	1,241,243,866	1,184,148,661	+4.8%	+13.1%
wikisort	1021	8482	833,317,333	26.65	32.89	1,057,727,371	1,187,236,518	−10.9%	+23.4%

The columns are numbered to enable referencing them

with (Column 6) integrating our core timing model, respectively. The simulation time denotes how long (wall time) it takes to run the benchmark on the VP. Please note that we have configured our core timing model to match the core of the HiFive1 board (see Sect. 3.2.1). The next two columns give the number of execution cycles for the benchmark as reported by the RISC-V VP with integrating our core timing model (Column 7, which provides a fast estimation of the real HiFive1 board) and the real HiFive1 board as comparison (Column 8). RISC-V provides a special register that holds the current number of execution cycles. We query this register before and after running the benchmark to obtain the spent execution cycles. Finally, the last two columns report the difference in the estimated number of execution cycles with the really measured execution cycles (Column 9) and the performance overhead in integrating our core timing model into the RISC-V VP (Column 10). Figure 3.16 presents the accuracy and performance overhead results in graphical form for a better illustration.

It can be observed that our core timing model provides a very accurate estimation of the number of execution cycles that we measured on the real HiFive1 board. For some benchmarks, our core timing model estimation under-approximates (i.e., estimates less cycles), and for other benchmarks over-approximates (i.e., estimates more cycles), the actual number of execution cycles. In summary, we observed a minimum and maximum difference of 0.0% and 13.5%, respectively, with an average of 4.7% on this benchmark set. These results demonstrate that our core timing model can be used for a very accurate timing/performance evaluation and hence enables an efficient VP-based design space exploration. The remaining accuracy gap is due to incomplete specification of the timing-related functionality of the HiFive1 board (e.g., the replacement strategy in the branch prediction buffers). We believe that with a more complete specification we can adapt the configuration of our timing model appropriately to obtain even more accurate results.

Furthermore, please note that our core timing model is not limited to the HiFive1 board but can be configured to match other specifications as well by configuring our pipeline, branch prediction, and caching models appropriately.

Besides high accuracy, our core timing model retains at the same time the high simulation performance of the VP by introducing a reasonably small performance overhead. We observed a minimum and maximum overhead of 10.0% and 37.4%, respectively, with an average of 18.9% on this benchmark set. With our core timing model, the RISC-V VP simulates with an average of 27 MIPS (*Million Instructions Per Second*) compared to 32 MIPS of the original RISC-V VP. The performance overhead depends on the number of data flow dependencies between subsequent instructions as well as number of branching instructions as they update our branch prediction model. These parameters are highly dependent on the actual benchmark. Furthermore, we have an additional (almost constant) time overhead for each instruction fetch, as we have to check (and potentially update) if the fetch results in a cache miss and hence incurs an execution time penalty.

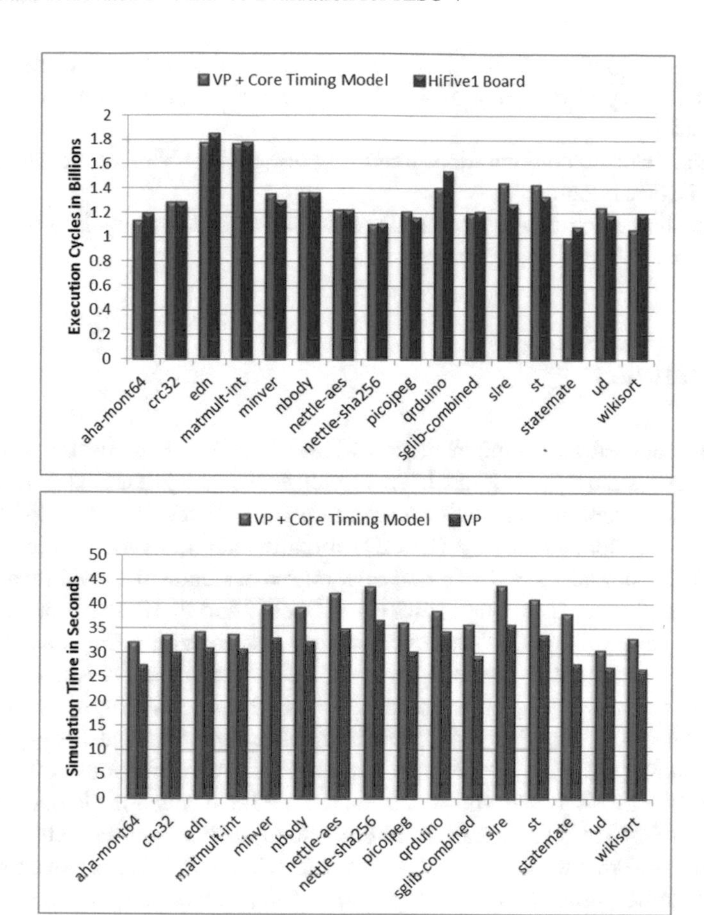

Fig. 3.16 Summary of the results on the accuracy compared to the HiFive1 board (top diagram) and performance overhead compared to the original RISC-V VP (bottom diagram) of our approach (VP + Core Timing Model)

3.2.4 Discussion and Future Work

Our proposed core timing model allows to consider pipelining, branch prediction as well as caching effects and showed very good results on a case study using the RISC-V HiFive1 board. In future work we plan to improve our work in the following directions:

- Investigate further abstractions of our timing models and generation of (e.g., basic block) summaries to simplify and reduce the number of model updates and hence reduce the performance overhead while retaining sufficient accuracy.

- Evaluate integration of DBT into the VP ISS for further performance optimization. This may also require adoptions of the timing model to ensure a smooth integration.
- Add and evaluate configurations matching other RISC-V cores, besides the core of the HiFive1 board.
- Extend the timing model to consider more advanced features like out-of-order execution.

3.3 Summary

We have proposed and implemented a RISC-V-based VP to further expand the RISC-V ecosystem. The VP has been implemented in SystemC and designed as extensible and configurable platform around a RISC-V RV32/64 (multi-)core with a generic bus system employing TLM 2.0 communication. In addition to the RISC-V core(s), we provide SW debug and coverage measurement capabilities, a set of essential peripherals, including the RISC-V CLINT and PLIC interrupt controllers, support for several operating systems (recently also including Linux support), and an example configuration matching the HiFive1 board from SiFive. The existing feature set in combination with the extensibility and configurability of our VP makes our VP suitable as foundation for various application areas and system-level use cases, including early SW development and analysis of interactions at the HW/SW interface of RISC-V-based embedded systems. Our evaluation demonstrated the quality and applicability to real-world embedded applications as well as the high simulation performance of our VP. Finally, our RISC-V VP is fully open source to stimulate further research and development of ESL methodologies.

In addition, we presented an efficient core timing model that allows to consider pipelining, branch prediction, and caching effects. We integrated the timing model into our open source RISC-V VP to enable fast and accurate performance evaluation for RISC-V-based systems. As a case study, we provided a timing configuration matching the RISC-V HiFive1 board from SiFive. Our experiments using the *Embench* benchmark suite demonstrate that our approach allows to obtain very accurate performance evaluation results (less than 5% mismatch on average) while still retaining a high simulation performance (less than 20% simulation performance overhead on average). To help in advancing the RISC-V ecosystem and stimulate research, our core timing model, including the integration with the RISCV VP, is available open source as well.

Chapter 4
Formal Verification of SystemC-Based Designs using Symbolic Simulation

In today's *Electronic System Level* (ESL) [12] design flow, *Virtual Prototypes* (VPs) play a very important role. They are essentially an abstract representation of the entire hardware platform and serve as a reference for embedded software development and hardware verification. For the creation of VPs, which are composed of abstract models of the individual IP blocks, the C++-based language SystemC [113] has been established as a standard. By virtue of their impact on the design flow, these abstract SystemC models should be thoroughly verified.

For this verification task, simulation-based approaches are still the main workhorse due to their scalability and ease of use. However, significant manual effort is required to create a comprehensive testsuite and in addition simulation-based methods cannot prove the absence of errors and often miss corner-case bugs. Formal verification methods are required to ensure the correctness. At the same time formal verification of high-level SystemC designs is a very challenging problem [212]. One has to deal with the full complexity of C++ to extract a suitable formal model (front-end problem) and then, with large cyclic state spaces defined by symbolic inputs and concurrent processes.

In Sect. 4.1 we present a scalable and efficient stateful symbolic simulation approach for SystemC that combines *State Subsumption Reduction* (SSR) with *Partial Order Reduction* (POR) and *Symbolic Execution* (SymEx) under the SystemC simulation semantics. While the SymEx+POR (i.e. symbolic simulation) combination provides basic capabilities to efficiently explore the state space, SSR prevents revisiting symbolic states and therefore makes the verification complete. The approach has been implemented on top of an *Intermediate Verification Language* (IVL) for SystemC to address the front-end problem. The scalability and efficiency of the implemented verifier are demonstrated using an extensive set of experiments. The approach has been published in [86, 100].

Building on the notable progress that our symbolic simulation technique has made on SystemC formal verification, we present a case study on applying it to

V. Herdt et al., *Enhanced Virtual Prototyping*,
https://doi.org/10.1007/978-3-030-54828-5_4

formally verify TLM peripheral models in Sect. 4.2. To the best of our knowledge, this is the first formal verification case study targeting this important class of VP components. First, we show how to bridge the gap between the industry-accepted modeling pattern for TLM peripheral models and the semantics currently supported by the SystemC IVL. Then, we report verification results for the interrupt controller of the LEON3-based SoCRocket VP used by the European Space Agency and reflect on our experiences and lessons learned in the process. This case study has been published in [135].

To further boost the scalability of symbolic simulation, we propose compiled symbolic simulation (CSS) in Sect. 4.3. CSS is a major enhancement that augments the DUV to integrate the symbolic execution engine and the POR based scheduler. Then, a C++ compiler is used to generate a native binary, whose execution performs exhaustive verification of the DUV. The whole state space of a DUV, which consists of all possible inputs and process schedules, can thus be exhaustively and at the same time very efficiently explored by avoiding interpreting the behavior of the DUV. An extensive experimental evaluation demonstrates the potential of our approach. The approach has been published in [88].

Finally, Sect. 4.4 presents an optimized CSS approach that integrates parallelization and architectural improvements to increase the scalability of CSS even further. The experiments demonstrate that parallelization can achieve significant performance improvements. This approach has been published in [90].

4.1 Stateful Symbolic Simulation

As already mentioned, to verify the abstract SystemC models, the straightforward approach is simulation offered already by the free event-driven simulation kernel shipped with the SystemC class library [1]. Substantial improvements have been proposed by supporting the validation of (TLM) assertions, see, e.g. [21, 48, 54, 204]. To further enhance simulation coverage, methods based on *Partial Order Reduction* (POR) have been proposed [20, 128]. They allow to explore all possible execution orders of SystemC processes (also referred to as *schedules*) for a given data input. However, representative inputs are still needed. Therefore, formal approaches for SystemC TLM have been devised. But due to its object-oriented nature and event-driven simulation semantics, formal verification of SystemC is very challenging [212]:

1. It must obviously consider all possible inputs of the *Design Under Verification* (DUV).
2. A typical high-level SystemC DUV consists of multiple asynchronous processes, whose different orders of execution (i.e. *schedules*) can lead to different behaviors, these must also be considered to the full extent by the verifier.

3. The defined state space very often contains cycles that arise naturally due to the use of unbounded loops inside the asynchronous processes.
4. The verifier is required to deal with the full complexity of C++ to extract a suitable formal model.

A promising direction to make the overall challenge more manageable is to work on an *Intermediate Verification Language* (IVL) that separates front-end and back-end issues. Such an IVL for SystemC has been introduced in our previous work [134]. The IVL is an open, compact, and readable language with a freely available parser designed for both manual and automatic transformations from SystemC. Moreover, all freely available benchmarks used by existing formal verification approaches for SystemC are also available in an extensive IVL benchmark set. We refer to the IVL technical report [92] for more background and detailed description of the IVL.

With the IVL in place one can focus on developing techniques to enhance the scalability and efficiency of the back-end (i.e. addressing the first three challenges). To this end, this book presents a stateful symbolic simulation approach together with a *State Subsumption Reduction* (SSR) technique at its heart, for the efficient verification of cyclic state spaces in high-level SystemC designs. More precisely, the stateful symbolic simulation combines SSR with POR and *Symbolic Execution* (SymEx) [122] under the SystemC simulation semantics. POR prunes redundant process scheduling sequences, while SymEx efficiently explores all conditional execution paths of each individual process in conjunction with symbolic inputs. Subsequently, the SymEx+POR combination enables to cover all possible inputs and scheduling sequences of the DUV exhaustively in a stateless manner. To deal with cycles, the stateful search keeps a record of already visited states to avoid re-exploration.

However, before discussing SSR, please let us note that two major challenges must be solved to enable a stateful search in combination with SymEx and POR. First, SymEx stores and manipulates symbolic expressions, which represent sets of concrete values. Therefore, the *state matching* process, required by a stateful search to decide whether a state has already been visited, involves non-trivial comparison of complex symbolic expressions. Second, a naive combination of POR with stateful search can potentially lead to unsoundness, i.e. assertion violations can be missed. This is due to the (transition) *ignoring problem*, which refers to a situation, where a relevant transition is not explored.

SSR solves the first challenge by applying *symbolic subsumption checking*, inspired by Anand et al. [10]. The basic idea is as follows: If the set of concrete states represented by a symbolic state s_2 contains the set of concrete states represented by a symbolic state s_1, s_1 is *subsumed* by s_2 and it is not necessary to explore s_1 if s_2 has already been explored. We employ a powerful exact subsumption checking method which involves solving a quantified SMT formula. As this computation is potentially very expensive, we also employ several optimizations. To address the second issue, we develop a tailored *cycle proviso* and show that its integration preserves the soundness of our stateful symbolic simulation (i.e. SSR+SymEx+POR combination).

We have implemented the techniques in a tool named SISSI (**S**ystemC **I**VL **S**ymbolic **Si**mulator). Our preliminary experiments [134] have already shown the potential of the SymEx+POR combination in comparison with all available state-of-the-art formal approaches. The experiments in this book focus on demonstrating the efficiency of the final stateful approach using an extensive set of benchmarks.

We start with a discussion of related techniques for formal verification of SystemC as well as intermediate representations related to our IVL and reduction techniques related to our SSR.

Related Formal Verification for SystemC

More recently, formal verification of RTL and HLS (i.e. high-level synthesizable) SystemC designs is supported by EDA tools such as OneSpin 360-DV or Calypto SLEC. Consequently, recent research shifts the focus to TLM, the abstraction level beyond HLS. SystemC TLM designs are characterized by communication via function calls and synchronization of asynchronous processes via events. They make very little or no use of signals and clocks. A handful of formal verification approaches for SystemC have been proposed in the past. Early efforts, for example, [82, 119, 155, 207], have very limited scalability or do not model the SystemC simulation semantics thoroughly [127]. Furthermore, they are mostly geared towards RTL signal-based communication. For a detailed overview of the early verification efforts, we refer to the related work section in [38]. In the following, we only focus on state-of-the-art approaches targeting TLM.

KRATOS [38] is the state-of-the-art SystemC model checker for handling cyclic state spaces. As input language, KRATOS accepts threaded C. The underlying model checking algorithm combines an explicit scheduler and symbolic lazy abstraction. POR is also integrated into the explicit scheduler to prune redundant schedules. For property specification, simple C assertions are supported. Although this approach is complete, its potentially slow abstraction refinements may become a performance bottleneck.

SCIVER [72] translates SystemC designs into sequential C models first. Temporal properties using an extension of PSL [205] can be formulated and integrated into the C model during generation. Then, C model checkers can be applied to check for assertion violations. High-level induction on the generated C model has been proposed to achieve completeness and efficiency. However, due to the absence of POR, it does not scale well to designs with a large number of processes.

SDSS [36] formalizes the semantics of SystemC designs in terms of Kripke structures. Then, BMC (and induction) can be applied in a similar manner as SCIVER. The main difference is that the scheduler is not involved in the encoding of SDSS. It is rather explicitly executed to generate an SMT formula that covers the whole state space. This approach has been extended further in [37] to include POR support. Still, the extended SDSS approach is incomplete and can only prove assertions in SystemC designs with acyclic state spaces.

STATE [84] translates SystemC designs to timed automata. With STATE it is not possible to verify properties on SystemC designs directly. Instead, they have to be formulated on the automata and can then be checked using the UPPAAL model checker. Our preliminary evaluation in [134] shows that STATE does not perform/scale well in the presence of symbolic inputs and also it does not accept many common SystemC benchmarks. Furthermore, STATE explores a much smaller state space in comparison to the other methods since its back-end does not support the full range of C++ *int*.

Related Intermediate Representations for SystemC

Threaded C, the input language of KRATOS, is the most close representation w.r.t. our IVL. However, it is not designed as an IVL and thus lacking documentation of the supported C subset and a freely available parser. Furthermore, SystemC-related constructs cannot be cleanly separated, e.g. processes and channel updates are identified by function name prefixes, events are declared as enum values, etc.

The RESCUE framework [81] is an ongoing effort to build a verification environment around STATE. RESCUE employs an intermediate representation called SysCIR and aims to provide various analysis and transformation engines (such as slicing and abstraction) on top of it. Conceptually, SysCIR is similar to our IVL; however, it has not been described beyond the concepts.

PinaVM [148] is the state-of-the-art SystemC front-end built on top of the LLVM compiler framework. PinaVM leverages LLVM to compile a SystemC design together with its customized SystemC library into the *LLVM Intermediate Representation* (LLVM-IR). Then, it employs JIT-compilation techniques to execute the SystemC elaboration phase to extract the architecture, i.e. available modules, processes, port bindings, etc., and stores it as *control flow graph* (CFG) in *static single assignment* (SSA) form with SystemC constructs. In principle, the output of PinaVM could be transformed further into our IVL and should be considered in future work.

Related Reduction Techniques

In the context of symbolic model checking for Java, a subsumption checking technique similar to ours has been considered [10]. However, the authors applied this technique to sequential Java programs, while we combine subsumption checking with POR under concurrency semantics of SystemC.

A dependency relation for POR techniques can be either statically or dynamically determined. Dynamic dependency is calculated during the simulation and thus is more precise than static dependency which is often an over-approximation. However, it produces a much bigger overhead in comparison to static dependency calculation. For a more detailed formal treatment of POR we refer to [57, 62, 128].

For the *ignoring problem* [210] a number of cycle provisos have been proposed as solution in combination with different search strategies, e.g. [53, 62]. However, in the context of SSR, these provisos are unsuitable, since they are too restrictive. This book presents a novel proviso for the combination of POR and SSR.

4.1.1 SystemC Intermediate Verification Language

The IVL is the stepping stone between a front-end and a back-end. Ideally, it should be compact and easily manageable but at the same time powerful enough to allow the translation of SystemC designs. Our view is that a back-end should focus purely on the behavior of the considered SystemC design. This behavior is fully captured by thc SystemC processes under the simulation semantics of the SystemC kernel. Therefore, a front-end should first perform the elaboration phase, i.e. determine the binding of ports and channels. Then it should extract and map the design behavior to the IVL, whose elements are detailed in the following.

Based on the simulation semantics described above, we identify the three basic components of the SystemC kernel: *SC_THREAD*, *sc_event*, and *channel update*. These are adopted to be *kernel primitives* of the IVL: *thread*, *event*, and *update*, respectively. Associated to them are the following primitive functions:

- *suspend* and *resume* to suspend and resume a thread, respectively;
- *wait* and *notify* to wait for and notify an event (the notification can be either immediate or delayed depending on the function arguments);
- *request_update* to request an update to be performed during the update phase.

These primitives form the backbone of the kernel. Other SystemC constructs such as *sc_signal*, *sc_mutex*, static sensitivity, etc. can be modeled using this backbone. The behavior of a *thread* or an *update* is defined by a function. Functions which are neither *threads* nor *updates* can also be declared. Every function possesses a body which is a list of statements. We allow only assignments, (conditional) goto statements, and function calls. Every structural control statement (*if-then-else*, *while-do*, *switch-case*, etc.) can be mapped to conditional *goto* statements (this task should also be performed by the front-end). Therefore, the representation of a function body as a list of statements is general and at the same time much more manageable for a back-end.

As *data primitives* the IVL supports Boolean and integer data types of C++ together with all arithmetic and logic operators. Furthermore, arrays and pointers of primitive types are also supported. Additionally, bit-vectors of finite width can be declared. This enables the modeling of SystemC data types such as *sc_int* or *sc_uint* in the IVL.

For verification purpose, the IVL provides *assert* and *assume*. More expressive temporal properties can be translated to FSMs and embedded into an IVL description by a front-end. The symbolic values of primitive types are also supported.

Example 1

(a) SystemC Example: Fig. 4.1 shows a simple SystemC example. The main purpose of the example is to demonstrate some elements of the IVL. The design has one module and three SC_THREADs A, B, and C. Thread A sets variable a to 0, if x is divisible by 2 and to 1 otherwise (Lines 13–16). Variable x is initialized with a random integer value on Line 6 (i.e. it models an input). Thread

```
1  SC_MODULE(Module) {
2    sc_core::sc_event e;
3    uint x, a, b;
4
5    SC_CTOR(Module)
6      : x(rand()), a(0), b(0) {
7      SC_THREAD(A);
8      SC_THREAD(B);
9      SC_THREAD(C);
10   }
11
12   void A() {
13     if (x % 2)
14       a = 1;
15     else
16       a = 0;
17   }
18
19   void B() {
20     e.wait();
21     b = x / 2;
22   }
23
24   void C() {
25     e.notify();
26   }
27 };
28
29 int sc_main() {
30   Module m("top");
31   sc_start();
32   assert(2 * m.b + m.a == m.x);
33   return 0;
34 }
```

Fig. 4.1 A SystemC example

B waits for the notification of event *e* and sets $b = x/2$ subsequently (Lines 20–21). Thread C performs an immediate notification of event *e* (Line 25). If thread B is not already waiting for it, the notification is lost. After the simulation the value of variable *a* and *b* should be x % 2 and $x/2$, respectively. Thus the assertion $(2 * b + a == x)$ is expected to hold (Line 32). Nevertheless, there exist counterexamples, for example, the scheduling sequence CAB leads to a violation of the assertion. The reason is that *b* has not been set correctly due to the lost notification.

(b) IVL Example: Fig. 4.2 depicts the same example in IVL. As can be seen the SystemC module is "unpacked," i.e. variables, functions, and threads of the module are now global declarations. The calls to *wait* and *notify* are directly mapped to statements of the same name. The *if-then-else* block of thread A is converted to a combination of conditional and unconditional `goto` statements (Lines 7–12). Variable *x* is initialized with a symbolic integer value (Line 2) and can have any value in the range of *unsigned int*. The statement *start* on Line 25 starts the simulation.

In short, the IVL is kept minimal but expressive enough for the purpose of formal verification. It covers all benchmarks used by existing formal verification approaches for SystemC. It would only take little effort to adapt these approaches to support this IVL as their input language. That would lead to the availability of a checker suite for SystemC once a capable front-end is fully developed. We also refer to an IVL description as a SystemC design since both define the same behavior. A grammar and a parser for the IVL are provided at our website http://www.systemc-verification.org/sissi together with an in-depth technical description [92].

```
1  event e;                          15 thread B {
2  uint x = ?(uint);                 16   wait (e);
3  uint a = 0;                       17   b = x / 2;
4  uint b = 0;                       18 }
5                                    19
6  thread A {                        20 thread C {
7   if x % 2 goto elseif             21   notify (e);
8     a = 0                          22 }
9     goto endif                     23
10  elseif:                          24 main {
11    a = 1                          25   start;
12  endif:                           26   assert (2 * b + a == x);
13 }                                 27 }
14
```

Fig. 4.2 The example in IVL

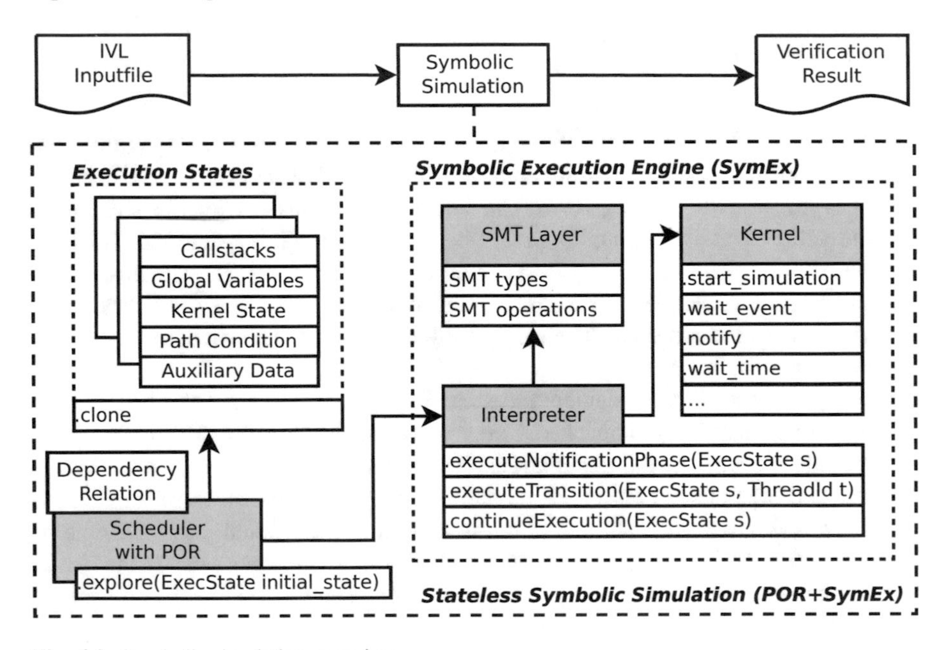

Fig. 4.3 Symbolic simulation overview

4.1.2 Overview Symbolic Simulation

Essentially, our stateless symbolic simulator consists of a (stateless) scheduler integrating POR and a SymEx engine. Figure 4.3 shows an overview. The scheduler (lower left side of Fig. 4.3) orchestrates the state space exploration. Therefore, it manages a frontier of reachable execution states. In each step the scheduler selects an active execution state s and explores a yet unexplored *runnable* thread t, i.e. s and

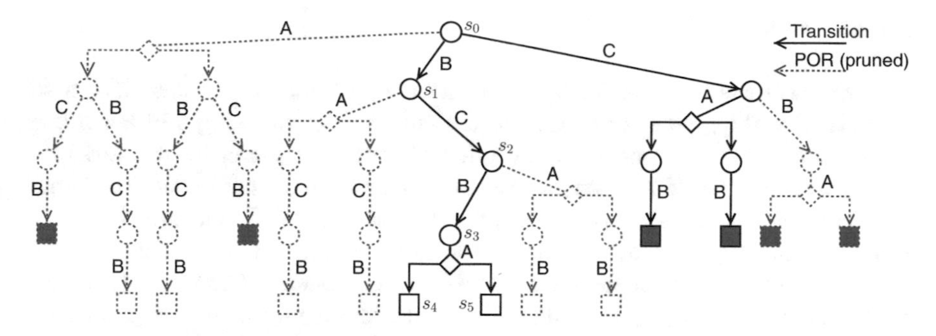

Fig. 4.4 Complete search tree for Example 1

t are passed to the SymEx engine for execution. The SymEx, shown on the right side of Fig. 4.3, will execute statements of t non-preemptively following the SystemC semantics, i.e. until a *wait* statement is reached, and update s accordingly resulting in a new state s_2. Such an execution is called a transition of thread t from state s to state s_2. Please note that the original state s is preserved in the scheduler by cloning, in order to explore different thread interleavings in s (however, s is discarded once all interleavings have been explored). Thus eventually all thread interleavings are explored for all execution states. POR is employed here in the scheduler to prune the exploration of redundant thread interleavings. We use a DFS-based (*Depth First Search*) scheduler; however, a different search strategy can be implemented in a similar way.

Example 2 Figure 4.4 shows the complete search tree (i.e. state space) for Example 1. As can be seen, even this very simple design has a total of 14 possible execution paths. Only six of them violate the assertion (paths that end with a filled box). The circles and boxes correspond to *execution states* of the design. An execution state is split at a $\diamond$ node in case of a conditional *goto* statement with both branch directions feasible (explained later in Sect. 4.1.2.2). The initial execution state s_0 has no incoming edges (the uppermost circle). It is constructed based on the IVL input file.

Every edge depicts a transition between two states and its label shows which thread has been selected and executed (by the symbolic execution engine) in this transition. From the initial state it is possible to start execution with either thread A, B, or C. Dashed lines denote redundant transitions which are pruned on the fly by applying POR techniques (explained later in Sect. 4.1.2.3) and thus do not have to be explored.

In the following we describe execution states, our SymEx engine, the employed POR method, and our stateless DFS-based scheduler in more detail.

4.1.2.1 Execution State

An execution state consists of a path condition PC, the kernel state KS, a set of callstacks, the global variables, and a Boolean flag indicating whether this is a *split* state. The path condition (PC) describes the constraint on the variables, which must be satisfied to reach this execution state from the initial one. The kernel state contains the status of each thread, the current simulation time and phase (e.g. evaluation phase), a list of pending notifications, and a pending update list. The status of a thread is either *runnable*, *blocked*, or *terminated*. The kernel state only contains concrete values. Every thread (and also the main function) is associated with a callstack to implement function calls. Callstacks consist of a list of (call-)frames, the first frame for the thread and another one for each active function call on top of it. Every such frame contains the local state of the function execution: an *instruction pointer* (IP) that determines the next statement to be executed, and a name to value mapping for local variables. Variables (local and global) in general have no concrete values, their values are rather expressions of symbolic and concrete values.

Example 3 Take the initial execution state s_0 in Fig. 4.4 (the uppermost node) as an example: The variable x is initialized as a symbolic value, while a and b have the initial value of zero. Each thread A, B, or C is *runnable* and has an IP pointing to the first statement of the thread body (Lines 7, 6, and 21 in Fig. 4.2, respectively). There are no pending event notifications, no pending updates, and yet no constraint on the path condition.

Now, the execution of thread B from s_0 will result in a new execution state s_1. The execution will change B's state to *blocked*, add e to the pending notification list, and update B's IP to Line 17. All other values remain unchanged.

4.1.2.2 Symbolic Execution Engine (SymEx)

The SymEx engine implements the symbolic execution of thread statements. It modifies the active execution state, which is selected by the scheduler. The interpreter component provides the interface to the scheduler. It keeps track of stack frames and global variables and supports folding of concrete values, i.e. 3 + 5 is simplified to 8 on the fly, as optimization. The SMT layer provides (bit-vector and array) types and operations to construct symbolic expressions, e.g. x + y. Kernel calls, i.e. *wait* and *notify*, are redirected to the kernel component for execution. In the following we will describe the execution of relevant instructions in more detail.

An assignment will simply overwrite the value of a variable with the right-hand side expression (rhs). The rhs can be either a concrete value or symbolic expression. Calls to the kernel are interpreted according to their semantics in SystemC, e.g. the *wait* function will block the thread execution, update the kernel state accordingly, and then return the control back to the scheduler. The verification functions *assume* and *assert* are interpreted according to their usual semantic: *assume(c)* will add c to the path condition ($PC = PC \wedge c$). If the new path condition is unsatisfiable, the execution path will not be further considered. An *assert(c)* call will check $PC \wedge \neg c$ or satisfiability using an SMT solver. If a solution is found, the assertion is violated, and a counterexample is created from this solution and reported. In addition to assertions, the SymEx engine also checks for other types of errors such as division by zero or memory access violation.

When executing a conditional goto (*if c goto L1*) with symbolic branch condition c in an execution state s, there are two possibilities: (1) both $PC \wedge c$ and $PC \wedge \neg c$ are satisfiable, i.e. both branch directions (the goto and fallthrough path) are feasible; (2) Only one of the branch directions is feasible. In this case the SymEx engine will simply continue with the feasible branch. In the first case s is split into two independent execution states s_T and s_F and their path condition (PC) is updated accordingly as $PC(s_T) := PC(s) \wedge c$ and $PC(s_F) := PC(s) \wedge \neg c$. Please note that only a single clone operation is necessary because the existing execution state s is reused, and the cloned state is accessible using *s.splitState*. The *split* flag is set for both states and the instruction pointer (IP) of s_T and s_F is set to continue at the label $L1$ (goto path) and the next instruction (fallthrough path), respectively. Finally, the control is returned to the scheduler. The scheduler decides which execution state to continue with first: s or *s.splitState*.

4.1.2.3 Partial Order Reduction (POR)

POR is employed in the scheduler to improve the scalability of symbolic simulation by pruning redundant thread schedules. Therefore, each thread is separated into multiple transitions. A transition corresponds to a list of statements that is executed non-preemptively following the SystemC semantics. Thus every (non-terminated) thread has a currently active transition, which is *runnable*, iff the thread is *runnable*. The first transition begins at the first statement of the thread. Subsequent transitions continue right after the context switch of the previous transition. In Example 1 thread A and C have a single transition, whereas thread B is separated into two transitions by the wait statement.

POR requires a dependency relation between transitions. Intuitively, two transitions t_1 and t_2 are dependent, if their execution does not commute, i.e. $t_1 t_2$ and $t_2 t_1$ lead to different results. In SystemC context, t_1 and t_2 are dependent if one of the following holds:

1. They access the same memory location identified by a variable, pointer, or array, with at least one write access;

2. One notifies an event immediately which the other transition waits for (e.g. thread B and C from Fig. 4.2);
3. A transition is suspended by the other.

Transition dependencies (obtained through a preliminary static analysis) are used at runtime to compute a subset of runnable transitions $r(s)$, called a *persistent set* [62] (see Definition 1), in each state s. Essentially, we employ the *stubborn-set* algorithm [62] to compute an arbitrary but deterministically defined persistent set for each state. Exploration of transitions, and hence threads, is limited to the persistent sets.

Definition 1 (Persistent Set) A set $r(s) = T$ of transitions enabled (i.e. runnable) in a state s is *persistent* in s iff, for all non-empty sequences of transitions

$$s = s_1 \xrightarrow{t_1} s_2 \xrightarrow{t_2} \dots s_n \xrightarrow{t_n} s_{n+1}$$

from s in A_G and including only transitions $t_i \notin T$, for $i \in \{1, \dots, n\}$, t_n is independent in s_n with all transitions in T.

Example 4 The dashed execution paths in Fig. 4.4 are pruned by using static POR and do not have to be traversed. For example, we have $r(s_0) = \{B, C\}$ and $r(s_1) = \{C\}$ in Fig. 4.4; thus, the execution of the runnable thread A is pruned from s_0 and s_1, respectively. With POR a reduced state space with 11 execution states is explored, compared to the complete state space with 36 states (please note that different persistent sets are possible, e.g. $r(s_0) = \{A\}$ instead of $r(s_0) = \{B, C\}$, but they will lead to similar reductions). This reduction can be intuitively explained as follows. Because A is independent of B and C, it is unimportant when A is executed, e.g. CAB, CBA, and ACB are equivalent. In contrast, the order between B and C is important due to their dependency.

4.1.2.4 Stateless Scheduler

Algorithm 1: Scheduler performing DFS-based symbolic simulation (SymEx+POR) for IVL

```
   Input: Initial State
1  states ← stack()
2  explore(initialState)
3  procedure explore(s) is
4    pushState(s)
5    while |states| > 0 do
6      s ← states.pop()
7      try
8        if s.expanded then
9          if |s.working| = 0 then
10           backtrack(s)
11         else
12           runStep(s)
13       else
14         expand(s)
15         runStep(s)
16     catch EndPath
17       pass
18 procedure backtrack(s) is
19   pass
20 procedure initialize(s) is
21   s.expanded ← False
22   s.working ← {}
23   s.done ← {}
24 procedure pushState(s) is
25   initialize(s)
26   states.push(s)
27 procedure expand(s) is
28   s.expanded ← True
29   if s.simPhase.isEval then
30     t ← selectElement(s.runnable)
31     s.working ← persistentSet(s, t)
32 function selectNextTransition(s) is
33   t ← selectElement(s.working)
34   s.working.remove(t)
35   s.done.add(t)
36   return t
37 procedure runStep(s) is
38   if s.simPhase.isEnd then
39     raise EndPath
40   else if s.simPhase.isEval then
41     n ← s.cloneState()
42     states.push(s)
43     t ← selectNextTransition(s)
44     executeTransition(n, t)
45   else if s.simPhase.isNotify then
46     n ← s.cloneState()
47     executeNotificationPhase(n)
48   else
49     continueExecution(s)
50     n ← s
51   if n.simPhase.isSplit then
52     pushState(n.splitState)
53   pushState(n)
```

A DFS-based (*Depth First Search*) exploration for the IVL is shown in Algorithm 1. Execution states are stored on a search stack. Every state s is associated with scheduling information (Lines 21–23): transitions already (*s.done*) and relevant to be explored (*s.working*) as well as a flag *s.expanded* to detect states that have been re-pushed on the search stack. The exploration algorithm loops in Lines 5–17 until the search stack is empty. In every iteration the next instructions of an execution state s are executed using the *runStep* procedure. In case s is a new state, it will be expanded first. The *expand* procedure computes a persistent set T in a state s starting with an arbitrary transition $t \in runnable(s)$ and stores T in *s.working* in the evaluation phase (*simPhase.isEval*). The behavior of *runStep* depends on the kernel simulation phase of the execution state s:

1. Will do nothing if the simulation has ended (*simPhase.isEnd*) and directly return to the main loop;
2. Explore one of the transitions t in *s.working* in the evaluation phase (*simPhase.isEval*)—t is executed on a new state n to preserve s for the other transitions;
3. Perform the other phases, i.e. update, delta, and timed notification (*simPhase.isNotify*). According to the SystemC simulation semantics, either n will reach the evaluation phase (new runnable transitions arise) or simulation end otherwise;
4. Simply continue the execution for a *split* state as they have been prepared during the split to continue with the true- (goto) or false- (fallthrough) branch, respectively.

In case the simulation has not ended, the resulting execution state n is added to the search stack (Line 53). In case of a *split* the state covering the other branch direction (*n.splitState*), which has been annotated by the interpreter during the split, is also collected on the stack (Line 52). Please note, for simplicity of presentation the *split* flag is encoded into the kernel simulation phase variable.

In the following we extend our stateless SymEx+POR combination to a stateful version by integrating SSR, a powerful algorithm for matching states with symbolic data.

4.1.3 State Subsumption Reduction

This section presents the main concepts of SSR in stateful symbolic simulation. We start with a motivating example that shows the benefits of SSR and demonstrates that SSR and POR are complementary.

4.1.3.1 Motivating Example

The IVL description in Fig. 4.5 consists of two threads: *increment* (I) and *guard* (G). They communicate through a global variable v and use the event e for synchronization. The *increment* thread increments v and then blocks until e is notified. The *guard* thread is scheduled to run once in every delta cycle. It ensures that v does not exceed the maximum value and performs a delta notification of

```
1  event e;                          14      notify e, 0;
2  int v = ?(int);                   15      wait_time 0;
3                                    16      if (v >= 2) {
4  thread increment {                17        v -= 1;
5    while (true) {                  18      }
6      wait e;                       19    }
7      v += 1;                       20  }
8    }                               21
9  }                                 22  main {
10                                   23    assume (0 <= v && v <= 2);
11 thread guard {                    24    start;
12   while (true) {                  25  }
13     assert (0 <= v && v <= 2);
```

Fig. 4.5 An IVL example

Table 4.1 Example data for the IVL example

	PC	v	$C(v)$	$runnable$
s_0	$x_1 \geq 0 \wedge x_1 \leq 2$	x_1	$\{0, 1, 2\}$	$\{I_1, G_1\}$
s_1	$x_1 \geq 0 \wedge x_1 \leq 2$	x_1	$\{0, 1, 2\}$	$\{G_1\}$
s_2	$x_1 \geq 0 \wedge x_1 \leq 2$	x_1	$\{0, 1, 2\}$	$\{I_2, G_2\}$
s_3	$x_1 \geq 0 \wedge x_1 \leq 2$	$x_1 + 1$	$\{1, 2, 3\}$	$\{G_2\}$
s_4	$x_1 \geq 0 \wedge x_1 \leq 2 \wedge x_1 + 1 < 2$	$x_1 + 1$	$\{1\}$	$\{I_2, G_2\}$
s_5	$x_1 \geq 0 \wedge x_1 \leq 2 \wedge x_1 + 1 \geq 2$	x_1	$\{1, 2\}$	$\{I_2, G_2\}$
s_6	$x_1 \geq 0 \wedge x_1 \leq 2 \wedge x_1 + 1 \geq 2$	$x_1 + 1$	$\{2, 3\}$	$\{G_2\}$
s_7	$x_1 \geq 0 \wedge x_1 \leq 2 \wedge x_1 + 1 \geq 2$	x_1	$\{1, 2\}$	$\{I_2, G_2\}$

the event e. In this example the maximum value is 2. Both threads consist of two transitions, denoted as I_1,I_2 and G_1,G_2, respectively, separated by the context switches in Lines 6 and 15. The whole simulation is unbounded and *safe*, i.e. the assertion in Line 13 always holds. For convenience we use high-level control flow structures, instead of (conditional) goto statements in this example, as they can be automatically transformed in a straightforward way.

A representative part of the complete state space is shown in Fig. 4.6. Circles represent states and edges depict transitions between them. A diamond represents a conditional branch, where both branches are feasible. The dashed triangles represent state space parts that are omitted to simplify the description. Initially, before the simulation starts (Line 24), v is assigned a symbolic integer literal x_1 in Line 2. Then v, and thus x_1, is constrained to the values $\{0, 1, 2\}$ in Line 23, resulting in the initial state s_0.

Now consider the transition sequence $I_1 G_1 I_2 G_2 I_2 G_2$. The relevant data of the involved states is shown in Table 4.1. It shows the path condition (PC), the symbolic expression representing variable v, the set $C(v)$ of all concrete values of v satisfying the path condition, and the set of runnable transitions.

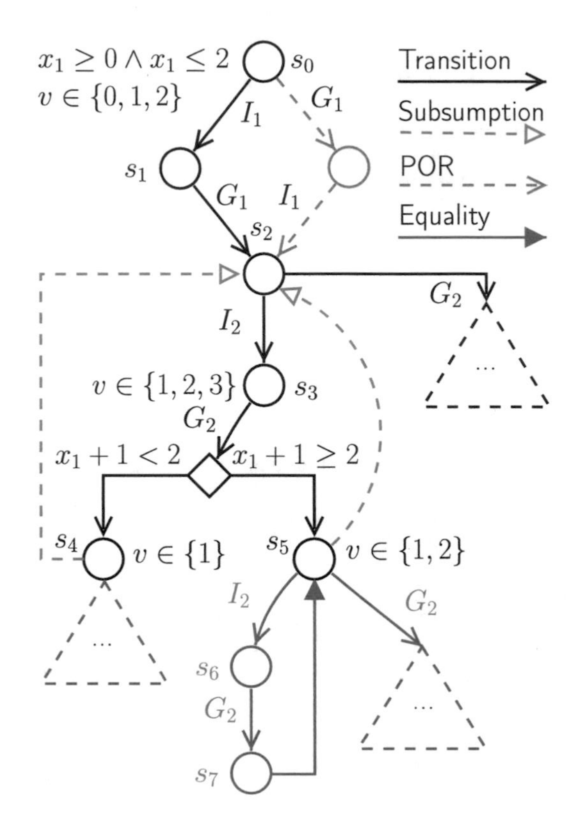

Fig. 4.6 State space for the IVL example

After I_1 is executed and thread *increment* is blocked by event e, s_1 is reached. Then, G_1 performs a delta notification of e, and after a delta notification phase, the state s_2 is reached, where both I_2 and G_2 become runnable.

Next, I_2 increments v to $x_1 + 1$ and reaches s_3. Then G_2 resumes from Line 16. Both, the branch condition $c_T = v \geq 2$ and its negation $c_F = v < 2$ are feasible in Line 16, with the current value of $v = x_1 + 1$ and path condition $x_1 \geq 0 \wedge x_1 \leq 2$. Thus the execution will fork two independent paths. The corresponding path conditions will be extended with c_T and c_F, respectively. In the path where c_T holds, v gets decremented in Line 17, whereas nothing happens in the other path. Eventually both paths will reach the context switch in Line 15, resulting in the states s_5 and s_4, respectively.

Execution of I_2G_2 from s_5 will reach the state s_6 and then s_7. Note that the execution of G_2 from s_6 does not fork at the conditional branch. The reason is that $v \in \{2, 3\}$ at this point; thus, the negated branch condition $v < 2$ is not satisfiable. The state s_7 is equal with s_5, as shown in Table 4.1; thus, s_5,s_6, and $s_7 = s_5$ form a cycle.

A *stateless search cannot detect the cycle* and would explore it infinitely unless the search is bounded. Conceptually, a stateful search, that is capable to detect the equality of $C(v)$ and *runnable* in s_5 and s_7, would solve the problem.

However, *much stronger reduction can be achieved by checking subsumption* of states. For example, the exploration of I_2G_2 from s_5 is actually unnecessary. The reason is $C(v)$ of s_5 is a subset of $C(v)$ of s_2 and the *runnable* sets are identical as shown in Table 4.1. Thus, any concrete states that are reachable from s_5 can also be reached from s_2. We say that s_5 is subsumed by s_2, and analogously, s_4 is also subsumed by s_2.[1] Thus the exploration of transitions from both s_4 and s_5 would be prevented by SSR. This is not possible with a simple stateful search based on equality checking. On large cyclic state spaces, SSR is expected to explore significantly less states.

Additionally, POR can be combined with SSR as a complementary reduction technique to further reduce the explored state space. In this example the execution of G_1I_1 from the initial state s_0 is pruned by POR, since I_1 has already been explored from s_0 and it is independent with any other transition.

However, as mentioned earlier, care must be taken to avoid unsoundness when applying POR in cyclic state spaces, and especially here in combination with SSR. In the following we introduce the concept of *weak reachability* and a cycle proviso based on this concept to solve the problem. The actual procedure for subsumption checking between two symbolic states is detailed in Sect. 4.1.4.

4.1.3.2 Weak Reachability

SSR results in the exploration of a reduced state space, denoted as A_R. Whenever it is detected that a state s_1 is subsumed by a state s_2, denoted as $s_1 \preccurlyeq s_2$, s_1 is not further explored. We say that there exists a *weak transition* from s_1 to s_2 in this case. This concept can be naturally extended to *weak reachability*. A state s' is *weakly reachable* from s, if it is reachable from s through a (possibly empty) sequence of normal or weak transitions. Intuitively when a state s' is reachable from a state s in the complete state space A_G, a state s'' will be weakly reachable from s in A_R, such that $s' \preccurlyeq s''$ holds. Thus, the reachability of concrete states and as a consequence, assertion violations are preserved through SSR.

Example 5 In the state space shown in Fig. 4.6, s_7 is reachable from s_5 by the sequence of transitions (trace) I_2G_2 in A_G. In A_R, s_5 is weakly reachable from itself by the same trace as follows: first s_2 is reached from s_5 by a weak transition and then s_5 is reached from s_2 by the trace I_2G_2. Since s_5 and s_7 are equivalent, $s_7 \preccurlyeq s_5$ holds.

[1] But neither s_4 nor s_5 is subsumed by any state from $\{s_0, s_1, s_3\}$ since they have different *runnable* transitions.

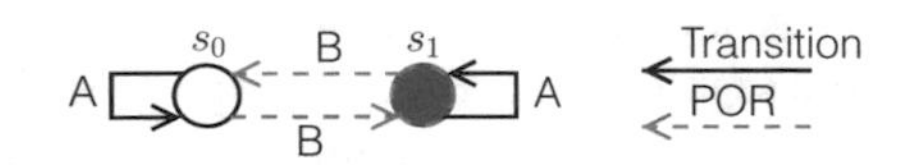

Fig. 4.7 Abstract state space demonstrating the (transition) ignoring problem

4.1.3.3 Cycle Proviso

A cycle proviso is required, to prevent transition ignoring when applying POR in the context of a stateful search and checking properties more elaborate than deadlocks. Otherwise, a relevant transition might be permanently ignored due to a cycle in the reduced state space.

Example 6 (Transition Ignoring Problem) Consider the state space in Fig. 4.7 with: two states s_0 and s_1; and two transitions A,B both runnable in both s_0 and s_1; and s_0 being the initial state. Execution of A will cycle in s_0 and s_1, and B will transition between s_0 and s_1. In case A and B are independent, $r(s_0) = \{A\}$ is a persistent set in s_0. Thus, only A is explored from s_0, i.e. execution of transition B is postponed since executions AB and BA are equivalent. However, a stateful search will stop exploration once the cycle in s_0 is detected. Therefore, transition B is never explored, i.e. is permanently ignored, and the error state s_1 is not reached.

To solve the ignoring problem, we have adapted the proviso C_2^S from [53], to the notion of weak reachability that arises in SSR, resulting in the C_{2W}^S proviso.

C_{2W}^S For every state s in A_R there exists a *weakly reachable* state s' from s in A_R, such that s' is fully expanded, i.e. every runnable transition in s' is explored.

In contrast the C_2^S proviso requires normal reachability of a fully expanded state, which would limit the reduction achieved by SSR. Recall the IVL example presented in Sect. 4.1.3.1. Since, e.g. $s_5 \preccurlyeq s_2$ holds, it is not necessary to further explore s_5; thus, no state is reachable from s_5 in A_R. However, the C_2^S proviso would enforce that a fully expanded state is reachable from s_5, resulting in the exploration of a larger state space. This inefficient behavior is solved with our C_{2W}^S proviso, as it only requires weak reachability.

A search algorithm that explores persistent sets and satisfies the C_{2W}^S proviso preserves assertion violations, as the following theorem states. However, the applicability of SSR and POR is not limited to the verification of safety properties.

Theorem 1 (Assertion Violation Preserving) *Let A_R be a POR and SSR reduced state space, which satisfies the cycle proviso C_{2W}^S. Let e_1 be an (arbitrary) error state and w be a sequence of transitions (trace) in A_G leading to e_1 from the initial state s_0. Then there exists an error state e_2 and a trace w_r in A_R such that e_2 is weakly reachable from s_0 in A_R.*

An error state is reached when an assertion violation is detected during transition execution. Please note that in the above theorem e_1 and e_2 are not required to be the same error state. For example, due to subsumption detection, the error state e_1 is pruned by the exploration, but instead another error state e_2 is reached with $e_1 \preccurlyeq e_2$. The above theorem can be shown by induction over the length of w. We provide a proof sketch in the following. For the proof we introduce another novel and weaker cycle proviso C_{2W}.

C_{2W} Let s be a state in A_R. Let $w = t_1..t_n$ be a non-empty trace leading to an error state from s in A_G. Then there exists a *weakly reachable* state s' from s in A_R, such that at least one transition t_i of w is in $r(s')$.

The C_{2W} proviso is specifically tailored to the requirements of the SSR algorithm. It ensures that some progress is made towards the execution of each trace. It allows to conduct the proof with minimal additional assumptions; therefore, it is very interesting from a theoretical perspective. Thus, whenever a new cycle proviso is introduced it is only necessary to show that it implies the C_{2W} proviso. This is exactly what we do for our C_{2W}^S proviso in the following lemma. Please note that in the definition of C_{2W}, transition t_1 is runnable in state s because t_1 is the first transition of a trace.

Lemma 1 (Cycle Proviso) *A POR and SSR reduced state space A_R that satisfies the cycle proviso C_{2W}^S also satisfies the cycle proviso C_{2W}.*

Proof The proof proceeds by induction and shows that every runnable transition is eventually explored when the cycle proviso C_{2W}^S is enforced.

Let s be an arbitrary state in a POR and SSR reduced state space A_R and t some transition runnable in s. According to C_{2W}^S a fully expanded state s' is weakly reachable by a trace w in A_R from s. Now either t is in $r(s_j)$ for some s_j visited when exploring w, or $t \in r(s')$.

Consider the first transition t_1 of w. In case t_1 is a weak transition, t is also runnable in the successor state by definition of $\preccurlyeq$. Otherwise, t_1 is a normal transition, there are two possibilities: (1) $t \in r(s)$, in which case the proof is complete or (2) $t \notin r(s)$, thus t is independent with all transitions in $r(s)$, since $r(s)$ is a persistent set in s. Therefore, t is still runnable in the successor state. By induction it follows that t will eventually be explored in A_R. □

A practical search algorithm will still implement the C_{2W}^S proviso because it can be efficiently integrated, e.g. in a DFS-based exploration algorithm. Conversely, it is not clear how to build a search algorithm that efficiently decides whether the weaker proviso C_{2W} is satisfied. Lemmas 1 and 2 together immediately imply Theorem 1.

Lemma 2 (Assertion Violation Preserving) *Let A_R be a POR and SSR reduced state space, which satisfies the cycle proviso C_{2W}. Let w be a sequence of transitions (trace) in A_G leading to an error state from the initial state s_0. Then there exists a trace w_r in A_R such that an error state is weakly reachable from s_0.*

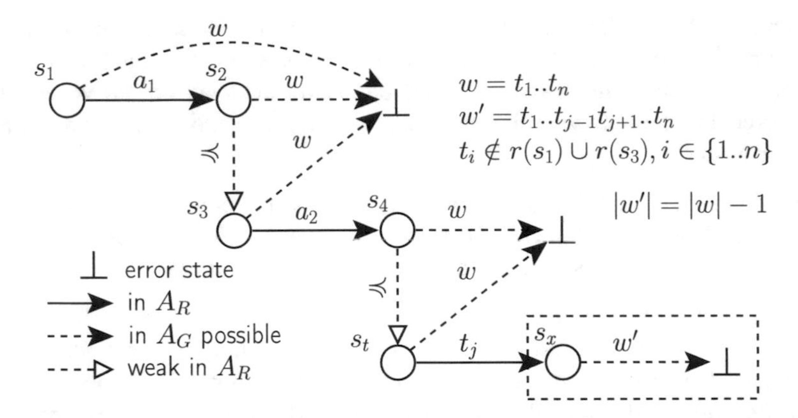

Fig. 4.8 Sketch of the proof idea for the main theorem

Proof The proof idea is illustrated in Fig. 4.8. It shows some arbitrary state s_1 from A_R (which can also be the initial state) and a trace $w = t_1..t_n$ that will lead to an error state in A_G when executed from s_1. Essentially, the reachability of the error state is preserved (1) when only executing transitions a_i, which are independent with transitions in w or (2) by weak transitions. For illustration Fig. 4.8 shows a chain of four transitions (two are weak) that preserve the reachability until a state s_t is reached. For example, executing a_1 from s_1 leads to a state s_2 in A_R. When no transition t_i is in the persistent set of s_1, then w will lead to an error state from s_2 too. Please note, in general this chain can be shorter ($s_t = s_1$ is also possible) or longer. However, the cycle proviso C_{2W} ensures that this chain is not infinite, i.e. some state s_y is weakly reached from s_1 in A_R such that some $t_i \in r(s_y)$. Wlog. let s_t be the first such s_y. Then, w will lead to an error state from s_t in A_G because the reachability of the error state is preserved in the chain.

Since $r(s_t)$ is a persistent set, executing the *smallest*[2] $t_j \in r(s_t)$ will lead to a state s_x such that $w' = t_1..t_{j-1}t_{j+1}..t_n$ will lead to an error state from s_x. The trace w' is shorter than w; otherwise, the same situation is valid for s_x as it was for s_1. Thus, the proof is completed by induction. □

[2] $t_a \notin r(s_t)$ with $a < j$, i.e. all t_a with $a < j$ are independent with t_b with $b \geq j$ in s_t.

4.1.3.4 Stateful Scheduler

Algorithm 2: Stateful symbolic simulation (SSR+ SymEx+POR) for IVL: extends Algorithm 1 by overriding the *expand*, *initialize*, and *backtrack* functions

```
Input: Initial State
1  H ← Set()
2  states ← stack()
3  explore(initialState)

4  procedure initialize(s) is
5    s.expanded ← False
6    s.working ← {}
7    s.done ← {}
8    s.safe ← False
9    s.unfinished ← False

10 procedure backtrack(s) is
11   if s.unfinished ∧ ¬s.safe then
12     refineWorkingSet(s)
13     states.push(s)

14 procedure markAllSafe(head) is
15   head.safe ← True
16   for s ∈ reversed(states) do
17     if s.safe then
18       break
19     s.safe ← True

20 procedure refineWorkingSet(s) is
21   s.working ← s.runnable \ s.done
22   markAllSafe(s)

23 procedure expand(s) is
24   s.expanded ← True
25   if s.simPhase.isEval ∨
       s.simPhase.isNotify then
26     if s ∈ H then
27       v ← H[s]
28       if v.safe then
29         markAllSafe(s)
30       else
31         v.unfinished ← True
32       raise EndPath
33     else
34       H.add(s)
35     if s.simPhase.isEval then
36       t ← selectElement(s.runnable)
37       s.working ← persistentSet(s, t)
38       if |s.working| = |s.runnable|
           then
39         markAllSafe(s)
40     else
41       markAllSafe(s)
```

Compared to the stateless search (see Algorithm 1 in Sect. 4.1.2.4), our stateful exploration (extensions shown in Algorithm 2) manages a set H of already visited states to avoid re-exploration and ensure that the cycle proviso C_{2W}^{S} is satisfied. The cycle proviso ensures that in the evaluation phase no relevant transition is permanently ignored due to cycles in the search space. Therefore, every state s is associated with two additional flags (Lines 8–9): s is marked *safe* when a fully expanded state is weakly reachable from s, and *unfinished* when s is detected on a cycle but not *safe* yet. The extensions in Algorithm 2 perform a subsumption based check to avoid re-exploration (Line 26, details on subsumption checking will be explained in Sect. 4.1.4) and ensure that every visited state is *safe*. A state s is discarded in one of the two cases (s will not be further considered by the search algorithm and thus needs to be *safe*):

1. s is backtracked once all transitions in $s.working$ have been explored. In case s is *unfinished* and not *safe*, i.e. s lies on a cycle of states and no other state on this cycle has been refined, s will be refined. Refinement simply ensures that s is fully expanded, i.e. all yet unexplored transitions in s are added to $s.working$. Thus, s and all states on the search stack, as they can reach s, become *safe*.
2. Another state v with $s \preccurlyeq v$ has already been visited (Line 27); thus, s is not further explored. In case v is safe, i.e. a fully expanded state is weakly reachable from v, s is also clearly safe. Otherwise, s and v lie on a cycle of states and v must still be on the search stack. In this case v is marked *unfinished*; therefore, it will be refined during backtracking if still necessary (v might become *safe* when a fully expanded state is reached before v is backtracked). In both cases s will also be *safe*.

In case a state is fully expanded it can immediately be marked *safe* (Lines 39 and 41, because all runnable transitions are explored or no transition is runnable at all, respectively). This is an optimization that can lead to less refined states during backtracking.

4.1.4 Symbolic Subsumption Checking

In this section we describe the actual procedure for subsumption checking between a new state s_1 and an already visited state s_2. For simplicity of the following representation, we assume that only global variables are used. The actual implementation supports local variables by matching stack frames. Thus, every (symbolic) value inside the program can be assigned a unique name. Let $V = \{v_1, \ldots, v_n\}$ be the set of all variables, the mapping *vars* of a state s can be denoted as $\{(v_1 : e_1^s), \ldots, (v_n : e_n^s)\}$, where e_i^s is the value of variable v_i in state s. We also refer to $SP(s) = \{(PC : PC(s), v_1 : e_1^s, \ldots, v_n : e_2^s)\}$ as the symbolic state parts of s. Please recall that KS(s) refers to the kernel state of s and only contains concrete values. It is thus included in the concrete state parts and is irrelevant for symbolic subsumption checking.

4.1.4.1 Exact Symbolic Subsumption (ESS)

Detecting that s_1 is subsumed by s_2 requires to show that the set of concrete states represented by a state s_1 is a subset of concrete states represented by s_2. A necessary condition for subsumption is thus $KS(s_1) = KS(s_2)$. Furthermore, if e_i^{s1} and e_i^{s2} are two concrete values, they must also be equal. Therefore, before trying subsumption on symbolic expressions, an equality test for these concrete state parts is performed. As an optimization we abstract away the current simulation time during this test, if the simulation is not bounded by time and the control flow of the program does not depend on the simulation time.

The subset condition for subsumption above can now be rephrased as follows: if a concrete state can be constructed from s_1 by assigning valid concrete values to its symbolic literals, then the same concrete state can also be constructed from s_2. The *Exact Symbolic Subsumption* (ESS) algorithm generates a quantified formula $F(s_1 \preccurlyeq s_2)$ that naturally encodes this requirement, as shown in the following:

$$\left(\exists x_1..x_p : PC(s_1) \wedge \bigwedge_{i \in \{1,\ldots,n\}} e_i^{s_1} = f_i\right) \implies \left(\exists y_1..y_q : PC(s_2) \wedge \bigwedge_{i \in \{1,\ldots,n\}} e_i^{s_2} = f_i\right)$$

Each term f_i is a fresh symbolic literal corresponding to the type of the variable v_i. The symbolic literals $x_1, \ldots x_p$ ($y_1, \ldots, y_q$) are all symbolic literals that appear in the variable values or the path condition of s_1 (s_2). In order to show that F is valid, we check its negation $\neg F$ for unsatisfiability. Any SMT solver with support for quantifiers can be employed.

Example 7 Consider $SP(s_1) = (PC\colon x_1 \neq 0,\ a\colon 2 * x_1,\ b : x_2)$ and $SP(s_2) = (PC\colon True,\ a\colon y_1 + 1,\ b\colon y_2 + y_3)$. All symbolic literals contained in s_1 and s_2 are $\{x_1, x_2\}$ and $\{y_1, y_2, y_3\}$, respectively. Two fresh symbolic literals f_1 and f_2 will be introduced with corresponding types for the variables a and b. The $\neg F(s_1 \preccurlyeq s_2)$ formula is as follows:

$$[\exists x_1, x_2 : (x_1 \neq 0) \wedge (2 * x_1 = f_1) \wedge (x_2 = f_2)] \wedge \neg [\exists y_1, y_2, y_3 : True \wedge (y_1 + 1 = f_1) \wedge (y_2 + y_3 = f_2)]$$

4.1.4.2 Optimizations

The ESS algorithm detects symbolic subsumption precisely, but it can be computationally very expensive due to the use of quantifiers. Therefore, we devise three optimization techniques:

1. *Explicit Structural Matching* (ESM) heuristically detects state equivalence (a special form of subsumption) by matching the structure of symbolic expressions;
2. *Filter Check* (FC) tries to refute subsumption by checking a simpler formula without quantifiers;

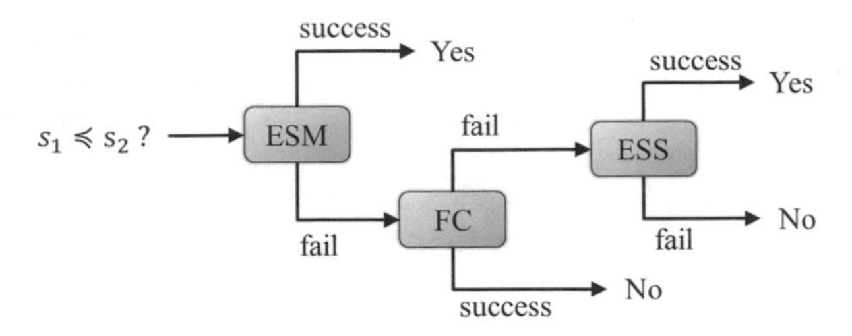

Fig. 4.9 Optimized ESS flow

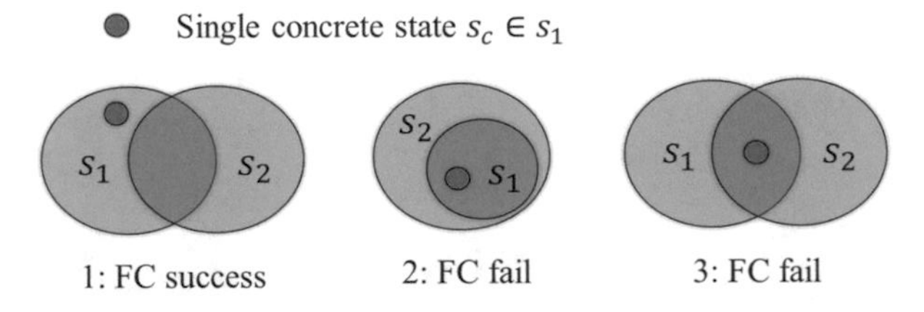

Fig. 4.10 Filter check principle

3. *Expression Simplification* (SIMP) is a generic technique that reduces the size of symbolic expressions and the path condition, thus also the size of the SMT formulas.

ESM and FC can in many cases avoid expensive ESS queries as shown in Fig. 4.9.

Explicit Structural Matching (ESM) The ESM heuristic is based on the simple observation that two symbolic expressions are semantically equal, if they are structurally equal. ESM checks whether every pair $(e_i^{s_1}, e_i^{s_2})$ as well as $(PC(s_1), PC(s_2))$ is equal except for the renaming of symbolic literals. Every symbolic literal has to be mapped to exactly another type-compatible symbolic literal. Matching the path conditions ensures that the mapped symbolic literals have the same constraints on both states. For example, ESM detects equivalence between $SP(s_1) = (pc\colon x_1 > 5, x_1 + 1)$ and $SP(s_2) = (pc\colon x_2 > 5, x_2 + 1)$ if x_1 and x_2 have the same type.

Filter Check (FC) The FC heuristic constructs a random concrete state s_C from s_1 and checks if s_C can be constructed from s_2. If the check fails, s_1 cannot be subsumed by s_2, and thus an ESS query is not necessary. This situation is shown on the left side of Fig. 4.10. Otherwise, s_1 might be subsumed by s_2, but not always as depicted on the right side.

The concrete state s_C is obtained by employing an SMT solver to solve $PC(s_1)$ and get a complete model (i.e. every symbolic variable is assigned to a concrete value). Such a model is always available, since the path to s_1 has been shown to be feasible before. Now, s_C can be constructed from s_2, iff the following formula is satisfiable:

$$PC(s_2) \wedge \bigwedge_{i \in \{1,\ldots,n\}} e_i^{s_2} = e_i^{s_C}$$

One implementation detail that is crucial to the overall performance of FC is to create and cache s_C when the feasibility of the path to s_1 is checked. This avoids an unnecessary solver query by FC to solve $PC(s_1)$ again.

Expression Simplification Symbolic expressions are simplified based on term rewriting, e.g. folding of concrete arguments ($1 + x + 2 \mapsto x + 3$) or simplification of special cases ($x \vee \neg x \mapsto T$).

The path condition is a conjunction of terms $PC = c_1 \wedge \ldots \wedge c_n$ representing constraints. During symbolic simulation, symbolic literals go out-of-scope when the variable using them is overwritten. For example, if x_1 is only used by the variable v, then x_1 goes out-of-scope when v is assigned to a new literal x_2. If a constraint c_i, which contains both out-of-scope and in-scope literals, does not exist, the out-of-scope literals can be safely removed. Constraints, which contain only these literals, also become irrelevant and thus are eliminated. This process is performed using standard garbage collection techniques.

4.1.5 Experiments

We have implemented the presented stateful symbolic simulation approach in Python (version 3.6.0) and evaluated it using the extensive set of benchmarks available in the IVL format. All experiments are performed on a 3.5 GHz Intel machine running Linux. The time and memory limits are set to 1000 s and 6 GB, respectively. The abbreviations T.O. and M.O. denote that the time and memory limit has been exceeded, respectively. N.S. denotes an unsupported benchmark. The result tables show the benchmark name in the first column and the verification result (V), with S for safe and U for unsafe, in the second column. All runtimes are specified in seconds. For unsafe benchmarks the verifier will stop once the first bug has been found, for safe benchmarks it needs to prove correctness. The Z3 solver [44] in version 4.5.0 is used to handle all symbolic queries, since it provides quantifier support as required for the ESS algorithm.

First, in Sect. 4.1.5.1 we show results for a comparison of the different optimization techniques, which have been discussed in Sect. 4.1.4.2, for our symbolic state matching algorithm ESS. Please recall that ESS is used for state matching in a stateful symbolic simulation (i.e. SSR+SymEx+POR). Then, we provide a detailed comparison of our approach with KRATOS [38], which is the state-of-the-art SystemC model checker for handling cyclic state spaces, in Sect. 4.1.5.2.

As a side note, the (sometimes significant) runtime improvements compared to the conference papers [86, 134] are due to a more recent version of Z3 and Python, minor improvements in SISSI's architecture and implementation, and a faster evaluation machine. This does not affect the gist of the experiments while gives a more up-to-date picture of SISSI.

4.1.5.1 Evaluation of ESS Optimizations

Table 4.2 shows the results of our evaluation of the optimization techniques for the ESS algorithm. The table is separated into two halves.

The left half shows the runtime in seconds for the base algorithm for subsumption checking (ESS), with structural matching (+ESM), with filter check (+FC), with expression simplifications (+SIMP) and a combination of all techniques (+ALL). All configurations use symbolic expression simplifications based on term rewriting. It can be observed that every optimization technique results in improvements compared to the base ESS algorithm. The combination of all optimization techniques yields the overall best results.

The right half provides additional information for the ESS+ALL configuration. It shows the observed factor of improvement (FoI) compared to the base algorithm (ESS), and how many state subsumption queries have been answered by the ESM (%ESM), FC (%FC) and ESS (%ESS) algorithm, respectively. For example, in the buffer benchmark all state subsumption queries can be answered using the ESM algorithm, thus FC and ESS are not used at all. For the *symbolic-counter* benchmark roughly 2 percent of the queries could already be resolved by the ESM algorithm and 87 percent are further passed to the FC algorithm, only 11 percent of all queries are decided by the ESS algorithm.

It can be observed that a substantial number (between 86% and 100%) of ESS queries can be avoided by employing our optimization techniques. In particular the FC test is very effective to refute subsumption. Furthermore, it can be observed that on some benchmarks the ESM algorithm is already sufficient for verification, as the ESS algorithm was not queried at all (%ESS=0). In case of safe benchmarks this means that the ESS algorithm was not necessary to detect re-exploration (and there has been no opportunity for subsumption based checking in the state space). For unsafe benchmarks a bug has been found before that was necessary. Furthermore, re-exploration of some state parts might be more efficient than employing the ESS algorithm. These results suggest to use ESM as a standalone heuristic state subsumption method. Therefore, we also incorporate the ESM algorithm in our following comparison.

4.1.5.2 Comparison with KRATOS

In this section we compare our optimized version of ESS against KRATOS the state-of-the-art model checker for SystemC. We also include the standalone ESM heuristic in the comparison as it has shown competitive results on many benchmarks. Furthermore, we add the stateless SymEx+POR in the comparison. It is unable

Table 4.2 Comparison of ESS optimizations (runtime in seconds)

Benchmark	V	ESS	ESS+ ESM	ESS+ FC	ESS+ SIMP	ESS+ ALL	FoI	%ESM	%FC	%ESS
buffer.p9	S	35.88	13.72	81.83	36.74	14.46	2.48	100	0	0
mem-slave-tlm-bug2.5	U	8.10	8.40	5.30	8.45	5.52	1.47	0	100	0
pressure2-sym.nb.50.5	S	233.01	229.95	157.34	229.64	158.38	1.47	0.62	89.14	10.23
pressure-sym.nb.50.5	S	86.49	87.60	70.24	88.49	70.60	1.23	0.47	85.62	13.91
symbolic-counter.1.15	S	6.01	6.09	5.04	6.10	5.00	1.20	2.18	87.05	10.76
token-ring2.12	S	101.20	94.34	61.05	88.10	39.00	2.60	15.20	84.80	0
token-ring-bug2.50	U	6.55	6.55	5.67	6.53	5.14	1.27	20.83	79.17	0
token-ring-bug.40	U	4.23	4.34	3.43	4.27	3.58	1.18	0	100	0
token-ring.40	S	67.23	68.04	53.99	67.58	48.57	1.38	8.85	91.15	0

The left half shows runtimes in seconds for the base algorithm (ESS, Sect. 4.1.4.1), with structural matching (+ESM, Sect. 4.1.4.2), with filter check (+FC, Sect. 4.1.4.2), with expression simplifications (+SIMP, Sect. 4.1.4.2) and a combination of all techniques (+ALL). The right half provides additional information for the ESS+ALL configuration. It shows the observed factor of improvement (FoI) compared to ESS, and how many state subsumption queries have been answered by the ESM (%ESM), FC (%FC), and ESS (%ESS) algorithm, respectively.

to prove the absence of errors in cyclic state spaces (these benchmarks are not supported and marked as N.S.) but can be effective in bug hunting.

Among the four state-of-the-art approaches mentioned in the related work section (Sect. 4.1), SDSS and its benchmarks are to the best of our knowledge not available. SCIVER does not provide POR techniques to prune redundant scheduling sequences. Therefore, SCIVER does not scale well to designs with a large number of processes. Our preliminary evaluation in [134] shows that STATE does not perform/scale well in the presence of symbolic inputs. Furthermore, STATE does not accept many common SystemC benchmarks and its back-end does not support the full range of C++ *int*. For these reasons, we do not consider SDSS, SCIVER, and STATE in our comparison.

The results of our evaluation are presented in Tables 4.3 and 4.4. In particular, Table 4.3 shows benchmarks with acyclic state spaces, whereas the more important cyclic state spaces are shown in Table 4.4. The third (*LOC IVL*) and fourth column (*#P*) show the lines of code in IVL and the number of processes, respectively. We have marked benchmarks originally proposed and used by KRATOS with a leading >. All runtimes are provided in seconds.

Our approach shows very competitive results compared to KRATOS. Improvements up to two orders of magnitude can be observed. This can especially be

Table 4.3 Comparison with KRATOS (1)—benchmarks with *acyclic* state spaces

		LOC			SSR+SymEx+POR		SymEx
Benchmark	V	IVL	#P	KRATOS	ESS+ALL	ESM	+POR
>kundu	S	64	3	0.29	1.04	0.92	1.07
>kundu-bug-1	U	48	2	0.18	0.46	0.46	0.47
>kundu-bug-2	U	64	3	0.16	0.58	0.58	0.53
>mem-slave-tlm2.1	S	197	2	0.57	1.13	1.12	1.13
>mem-slave-tlm2.5	S	233	2	17.20	20.46	10.19	12.86
>mem-slave-tlm-bug2.1	U	195	2	0.38	1.11	1.10	1.10
>mem-slave-tlm-bug2.5	U	233	2	10.93	5.52	2.37	1.39
>mem-slave-tlm-bug.1	U	183	2	0.67	0.99	0.99	0.99
>mem-slave-tlm-bug.5	U	219	2	50.85	1.24	1.23	1.21
>mem-slave-tlm.1	S	183	2	0.88	0.99	1.10	1.06
>mem-slave-tlm.5	S	219	2	66.17	1.18	1.20	1.19
pressure-safe.13	S	31	3	0.74	0.47	0.47	22.22
pressure-safe.100	S	31	3	T.O.	1.18	1.15	T.O.
pressure-sym.10.5	S	33	3	0.51	0.87	0.85	75.73
pressure-sym.50.5	S	33	3	67.24	2.43	2.30	T.O.
buffer-ws.p4	S	69	5	0.72	0.71	0.70	4.45
buffer-ws.p8	S	101	9	36.91	6.47	6.43	T.O.

Runtime provided in seconds. Benchmarks from KRATOS are marked with >. KRATOS is called with the options: `-opt_cfa 2 -inline_threaded_function 1 -dfs_complete_tree=false -thread_expand=DFS -node_expand=BFS -po_reduce -po_reduce_sleep -arf_refinement=RestartForest`

Table 4.4 Comparison with KRATOS (2)—benchmarks with *cyclic* state spaces

		LOC			SSR+SymEx+POR		SymEx
Benchmark	V	IVL	#P	KRATOS	ESS+ALL	ESM	+POR
buffer.p8	S	99	9	23.68	6.49	6.49	N.S.
buffer.p9	S	107	10	526.78	14.46	14.30	N.S.
pc-sfifo-sym-1	S	49	2	0.14	0.50	T.O.	N.S.
pc-sfifo-sym-2	S	65	2	0.17	0.53	T.O.	N.S.
pressure2-sym.nb.10.5	S	52	5	1.25*	12.23	T.O.	N.S.
pressure2-sym.nb.50.5	S	52	5	153.30*	158.38	T.O.	N.S.
pressure-sym.nb.10.5	S	33	3	0.48*	5.58	3.66	N.S.
pressure-sym.nb.50.5	S	33	3	51.90*	70.60	5.22	N.S.
rbuf-2	S	82	3	15.58*	0.67	T.O.	N.S.
rbuf-4	S	108	5	16.04*	0.79	T.O.	N.S.
simple-fifo-1c2p.20	S	83	3	37.03	1.59	1.58	N.S.
simple-fifo-1c2p.50	S	83	3	T.O.	3.99	4.00	N.S.
simple-fifo-2c1p.20	S	83	3	33.47	2.23	2.19	N.S.
simple-fifo-2c1p.50	S	83	3	T.O.	6.07	6.09	N.S.
simple-fifo-bug-1c2p.20	U	79	3	18.28	1.06	1.06	0.82
simple-fifo-bug-2c1p.20	U	79	3	14.68	1.17	1.18	0.88
symbolic-counter.1.15	S	41	3	559.96	5.00	2.08	N.S.
symbolic-counter.9.15	S	41	3	155.02	2.06	284.44	N.S.
>token-ring2.12	S	224	13	42.62	39.00	10.17	N.S.
>token-ring2.20	S	360	21	M.O	229.43	47.17	N.S.
>token-ring-bug2.17	U	308	18	22.84	1.76	1.58	1.55
>token-ring-bug2.50	U	869	51	M.O	5.14	4.68	4.78
>token-ring-bug.20	U	220	21	133.20	1.95	0.94	0.78
>token-ring-bug.40	U	420	41	M.O	3.58	1.89	1.36
>token-ring.13	S	150	14	2.02	2.55	1.06	N.S.
>token-ring.15	S	170	16	32.09	3.20	1.40	N.S.
>token-ring.40	S	420	41	M.O	48.57	10.16	N.S.
toy-sym	S	86	5	0.46	1.56	T.O.	N.S.
>transmitter.50	U	414	51	1.13	1.60	1.58	N.S.
>transmitter.90	U	734	91	M.O	3.54	3.52	N.S.
>transmitter.100	U	814	101	65.69	4.21	4.20	N.S.

Runtime provided in seconds. Benchmarks from KRATOS are marked with >. KRATOS is called with the options: `-opt_cfa 2 -inline_threaded_function 1 -dfs_complete_tree=false -thread_expand=DFS -node_expand=BFS -po_reduce -po_reduce_sleep -arf_refinement=RestartForest`

observed with up-scaled benchmarks, e.g. the *token-ring*, *mem-slave-tlm*, *transmitter*, and *pressure* benchmarks. This also demonstrates the scalability of our approach. For example, (nearly) doubling the number of processes from 51 to 101 in the transmitter benchmark does increase the runtime of our tool by 2.6x, whereas the runtime of KRATOS is increased by 58.1x. On some benchmarks, KRATOS shows better results but the runtime differences are not significant. This can be explained as follows. KRATOS starts with a coarse abstraction of the design and gradually refines it until a (real) counterexample is detected or safety is proven on the (conservative) abstraction. This is a complete approach that works very well for benchmarks that require only a small number of refinement steps. Furthermore, our approach is implemented in an interpreted language and thus might have a larger overhead on easy benchmarks. Also, on some safe benchmarks, KRATOS reports spurious counterexamples. These benchmarks are marked with *.

The standalone ESM heuristic can further speed up the verification process. The reason is that the ESS query is quite expensive and thus adds additional runtime overhead if no significant state subsumption is detected within the state space. This can especially be observed for the token-ring family of benchmarks. However, the experiments also clearly demonstrate that state subsumption checking can be very important to reduce the explored state space. Much stronger reduction can be achieved by checking the subsumption of states compared to equality when symbolic values with complex expressions are involved.

Furthermore, stateful symbolic simulation avoids redundant re-exploration of already visited states; thus, it can also improve the performance when applied to acyclic state spaces. This can, for example, be observed on the pressure benchmarks when comparing the runtime of our stateless approach (SymEx+POR) with our stateful approach (ESM, ESS). While SymEx+POR is very effective in bug hunting, as the experiments demonstrate, it is less effective in proving correctness as it does not detect redundant states.

The state space of the transmitter benchmark contains closed cycles from which no path to an error state exists. Therefore, it is not supported by the stateless SymEx+POR as the scheduler is DFS-based and thus can get stuck in a cycle without an error state. This problem could however be solved by using a randomized state space exploration algorithm. We expect it to perform similarly as in the *token-ring-bug* benchmark then.

4.1.6 Discussion and Future Work

Our stateful symbolic simulation currently does not support state merging, employs only static POR, and stores all observed execution states. While the experiments already demonstrated the efficiency of our approach on the available high-level SystemC benchmarks from the literature, the scalability of our approach can be further improved, by incorporating the above-mentioned techniques, in order to tackle verification of real-world virtual platforms. We discuss the advantages, drawbacks, and implications of doing so in the following.

4.1.6.1 Path Merging

In general our approach immediately works together with path merging. However, our first preliminary experiments with path merging showed that it can further speed up a stateful symbolic simulation but at the same time can have negative performance effects. We observed both cases on our benchmark set. On most benchmarks, however, path merging showed no significant effect. The reason is twofold: (1) ESS is a powerful state matching algorithm that can prune execution paths early and (2) our merging algorithm does not support the merging of the kernel state as it is implemented using concrete data structures. This results in less opportunities for merging.

In general a path merging algorithm should balance between complex symbolic expressions and number of paths explored. Otherwise, the benefit of exploring less execution paths is negated by too expensive solver queries. Due to quantifiers this negative effect is further amplified with the ESS algorithm. Furthermore, a naive combination of path merging with SSR seems not very efficient. The reason is that SSR detects subsumption between whole states, but by using path merging, an execution state is naturally partitioned into multiple parts. It seems more useful to adapt the subsumption checking to incorporate merging specific knowledge and thus match state parts with each other. This allows to simplify the complex symbolic expression due to state merging.

Thus, we leave it for future work to combine path merging with our stateful symbolic simulation in an efficient manner.

4.1.6.2 Dynamic Partial Order Reduction

Dynamic POR calculates dependencies during the simulation and thus more precisely than static POR. DPOR can improve scalability especially when shared pointers and arrays are involved. We believe that the increased precision is worth the additional overhead involved in computing dependencies during the simulation. However, it is not immediately apparent if the combination of SSR with DPOR is sound. We leave it for future work to investigate this combination.

4.1.6.3 State Matching Heuristics

Our experiments showed that using ESM alone as heuristic state matching algorithm, without relying on the ESS algorithm, can speed up the verification process. The reason is that the re-exploration of some state space part might be faster than the runtime cost for the ESS queries. Therefore it seems promising to: (1) Devise different state matching heuristics that balance between precision and runtime cost; (2) Dynamically decide at runtime which algorithm to use for every pair of states. The reason is that some checks are more important than others. An undetected cycle will lead to non-termination of the verification process; thus, checks with states on the search stack are more important; (3) Set an (adaptive) time limit for the ESS algorithm to avoid getting stuck on complex problem instances. The time limit can be gradually increased for states on the stack and adjusted individually for other states. A simple heuristic might associate each state with the (expected) time required to fully explore the state space reachable from it, e.g. take a timestamp when the state is entered and left.

4.2 Formal Verification of an Interrupt Controller

In recent years, notable progress on SystemC formal verification has been made. From the first approaches (about 15 years back), which only supported RTL-like models (e.g. [47, 70]), a new generation of SystemC verifiers [20, 36, 38, 72, 83, 100] has emerged to cope with the abstract modeling styles of SystemC TLM, as discussed in the previous section. In this context, an extensive set of academic SystemC benchmarks is available. From the practical perspective, however, it is mandatory to investigate the applicability of existing formal approaches on real-world VPs.

A first attempt has been made in [172], where the successful application of [83] on a simplified ARM AHB TLM-2.0 model is reported. Besides such memory mapped buses that are modeled using the TLM-2.0 standard [113], TLM peripherals (i.e. TLM models of on-chip peripherals such as timer, interrupt controller, memory controller, ADC, etc.) occupy an essential portion of real-world VPs. For this class of models, the separation of communication, behavior, and storage in a reusable way has also come into sharp focus [123]. This resulted in specialized modeling concepts for memory mapped register access and abstract wire communication (for interrupts and reset) at TLM. These concepts have been widely embraced and implemented in several industrial solutions, e.g. the *SystemC Modeling Library* (SCML) by Synopsys [203] or the joint proposal of TLM extensions by Cadence and ST [202], as well as in the open-source offer GreenReg [66], which is essentially a refinement of Intel's *Device Register Framework* (DRF).

In this work, we report preliminary results on formal verification of TLM peripheral models, which is part of a long-term effort towards a formally verified VP. The main contribution is a feasibility case study involving a real-world TLM model of an interrupt controller. The model is taken from the recently available open-source VP SoCRocket [189], which represents the popular LEON3 SoC platform [68] and has been developed in a joint effort with the *European Space Agency* (ESA). In order to achieve the results, a new modeling layer has been created to handle the above-mentioned modeling constructs, i.e. in particular register and wire modeling at TLM. The modeling layer and the models are available at www.systemc-verification.org/sissi.

```
1 void Irqmp::init_registers() {
2   regfile.create_register("pending", // register name
3     "Interrupt Pending Register", // description
4     0x04, // address offset
5     0x00000000, // initial value
6     IR_PENDING_EIP | IR_PENDING_IP) // write mask
7     .callback(POST_WRITE, this, &Irqmp::pending_write);
8   // ...
9 }
```

Fig. 4.11 TLM register modeling example

4.2.1 TLM Peripheral Modeling in SystemC

This section gives a compact overview on two industry-accepted modeling concepts for TLM peripherals: register and wire modeling. We use the particular implementation within SocRocket, which implements GreenReg-like API on top of the above mentioned extensions by Cadence and ST, as demonstration.

4.2.1.1 TLM Register Modeling

In practice, for TLM peripherals, memory mapped (control) register access is very common. Hence, the design productivity can be improved by greatly simplifying the description of registers and providing a reusable API for typical register features. Now consider Fig. 4.11 as an example. First, the member function *create_register* of the register file class is called to create a new register in Line 2. The arguments are the short register name, the full description (Lines 2–3), the address offset (Line 4), the initial value (Line 5), and the write mask (Line 6). The address offset allows to map read/write TLM-2.0 bus transaction to the register. The write mask, represented by the ORed value of two integer constants, determines which bits are allowed to be written by such a transaction.

After its creation, behavior is attached to the register by adding *function callback* (Line 7). Whenever the register is accessed via a TLM transaction, the callback is invoked. The invocation can be before/after reading/writing. In the example, the function *pending_write* of the *Irqmp* class will be called after the register value is updated (i.e. *POST_WRITE*). Furthermore, in a similar way, callbacks can also be registered on value changes of a single bit or a group of bits instead of the whole register. For the simplicity of presentation, we only show the latter.

4.2.1.2 TLM Wire Modeling

A major interoperability problem occurs in practice, since the TLM-2.0 standard does not specify how TLM wires are modeled. TLM wires are needed in particular for interrupt and reset signals where standard transaction payloads are too costly from both modeling perspective and simulation performance. With the naive solution of using *sc_signal* of SystemC, value updates require context switches in SystemC kernel. In the proposal by Cadence and ST (also in discussion for standardization within Accellera), value updates to TLM wires are immediately available (thus sidestepping the SystemC kernel) and a callback can be registered to execute desired behavior on value changes in a similar manner to TLM register callbacks.

4.2.2 Bridging the Modeling Gap

An analysis of the existing SystemC formal verification approaches reveals that none of them directly supports the described TLM peripheral modeling concepts. Developing a capable SystemC front-end, which additionally handles the TLM register and wire modeling constructs, would be very daunting. The other alternative is to map these constructs into intermediate representations supported by existing verification approaches. These representations, such as the threaded C subset supported by Cimatti et al. [38] or the *Intermediate Verification Language* (IVL) proposed in [134], however, contain only primitive imperative programming constructs. A manual translation of the abstract peripheral modeling constructs into these would require a considerable amount of very error-prone work.

Conversely, TLM peripheral modeling constructs are implemented as C++ classes and thus, they can be naturally modeled using *object-oriented* (OO) constructs. Therefore, our solution is to extend the open-source available IVL with minimal OO support, just enough to capture the behavior of TLM registers, wires, and callbacks. To be more specific, we have added support for classes, inheritance, virtual methods with overrides and dynamic dispatch. We refer to this extended version of IVL as XIVL in the remainder of this book.

Each TLM register or wire is modeled as an XIVL class with various member variables such as its values, associated callbacks, etc. A generic callback is modeled as an IVL class with an abstract member function, which should be implemented by a specific callback subclass to add the desired behavior. The register file of a TLM peripheral is modeled as an array of registers. In the following, we discuss the modeling of TLM registers and callbacks in XIVL in more detail. The modeling of TLM wires in XIVL is analogous and thus omitted for a more compact presentation.

An XIVL implementation of the generic *Register* class is shown in Fig. 4.12. The *bus_read/write* functions mimic a TLM read/write transaction to the register, since they will additionally invoke registered callbacks as shown in Lines 15 and 18, respectively. The *add_callback* function shown in Line 21 is used to register a generic callback object that adheres to the *RegCallback* interface shown in Line 1. A concrete callback implementation is shown in Fig. 4.13. On construction, the receiver class is stored as a member of the *PendingWriteCallback* class, as shown in the construction function in Line 9. This allows to delegate the callback to the

```
1  struct RegCallback {
2    abstract void call(RegCallback *this);
3  }
4  #define PRE_WRITE 0
5  ...
6  struct Register {
7    uint32_t value;
8    uint32_t write_mask;
9    RegCallback *callbacks[MAX_CALLBACKS];
10   void write(Register *p, uint32_t value) {
11     p->value = value;
12   }
13   void bus_write(Register *p, uint32_t value) {
14     if (p->callbacks[PRE_WRITE] != nullptr)
15       ::call(p->callbacks[PRE_WRITE]);
16     ::write(p, value & p->write_mask);
17     if (p->callbacks[POST_WRITE] != nullptr)
18       ::call(p->callbacks[POST_WRITE]);
19   }
20   // ...set_bit, get_bit, read, bus_read...
21   void add_callback(Register *p, uint32_t index,
22         RegCallback *c) {
23     p->callbacks[index] = c;
24   }
25 }
```

Fig. 4.12 Register class in XIVL

```
1  struct PendingWriteCallback extends RegCallback {
2    Irqmp *receiver;
3    virtual void call(PendingWriteCallback *this) {
4      ::pending_write(this->receiver);
5    }
6  }
7  RegCallback *create_pending_write_callback(Irqmp *p) {
8    PendingWriteCallback *c = new PendingWriteCallback;
9    c->receiver = p;
10   return c;
11 }
```

Fig. 4.13 Example callback implementation in XIVL

corresponding function of the receiver in Line 4, in this case the *pending_write* function of the *Irqmp* class. Basically, the code fragment in Fig. 4.13 is a template that is replicated, while adapting types and arguments, for every required callback. Finally, Fig. 4.14 shows an XIVL fragment, which corresponds to the SystemC code in Fig. 4.11, to create and initialize a register. The *create_register* function (implementation not shown in Fig. 4.12) allocates and initializes a new register with the given initial value and write mask. The resulting register is stored in the register file array *regfile*, which is a member of the *Irqmp* class. The initial value and write mask are associated with the register, whereas the address offset corresponds to the array index in *regfile*.

```
1  void init_registers(Irqmp *p) {
2    RegCallback *c = create_pending_write_callback(p);
3    p->regfile[0x04] = create_register( // 0x04=address off.
4      0x00000000, // initial value
5      IR_PENDING_EIP | IR_PENDING_IP); // write mask
6    ::add_callback(p->r[0x04], POST_WRITE, c);
7    // ...
8  }
```

Fig. 4.14 Register initialization in XIVL

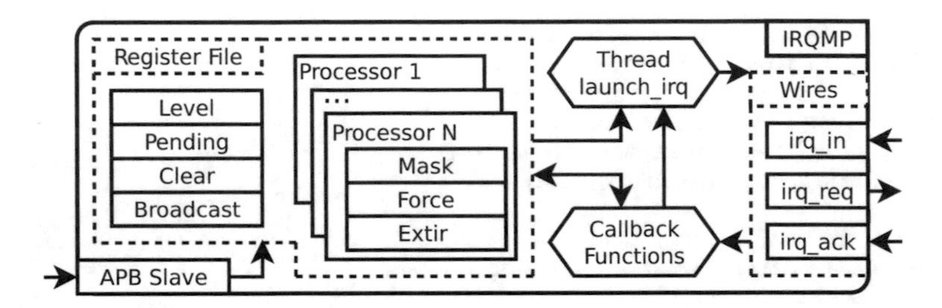

Fig. 4.15 IRQMP overview

On a final note, XIVL mostly resembles C++, except for some following minor differences: (1) the *extends* keyword is used for inheritance; (2) the implicit first argument *this* to member functions is declared and passed explicitly; (3) member functions are called by prepending *::* before the function name, which allows to distinguish between static and potentially dynamic calls.

4.2.3 *Case Study*

This section presents our effort to formally verify the TLM model of the Interrupt Controller for Multiple Processors (IRQMP) of the LEON3-based VP SoCRocket [189]. Thanks to the added modeling layer, we were able to translate the original SystemC TLM model into XIVL rather effortlessly. In the following first the basics of the IRQMP are described. Then, the obtained verification results are reported.

4.2.3.1 Interrupt Controller for Multiple Processors

The IRQMP is an interrupt controller supporting up to 16 processors. Its functionality is to process incoming interrupts by applying masking and prioritization of interrupts for every processor. Figure 4.15 shows an overview of the TLM model. It consists of a register file, several input and output wires, and an APB Slave interface.

The register file and wires are implemented as described in Sect. 4.2.1. The SystemC thread *launch_irq* and the callbacks implement the behavior logic of the IRQMP. The callbacks can update register values and activate the *launch_irq* thread, which is responsible to compute the interrupt requests for each processor.

In the following, we briefly summarize the specified behavior of the IRQMP described in the specification from [68]. IRQMP supports incoming interrupts (*irq_in*) numbered from 1 to 31 (interrupt line 0 is reserved/unused). Lines 15:1 are regular interrupts and Lines 31:16 are extended interrupts. In regular operation mode, IRQMP ignores all incoming extended interrupts. The *irq_req* and *irq_ack* wires are connected with every processor and allow to send interrupt requests and receive acknowledgments. The register file contains shared and processor-specific registers. Every register has a bit width of 32. Each bit naturally represents an interrupt line.

Interrupt Handling IRQMP computes interrupt requests for every processor P by first, combining Pending with Force and Mask register of P using bitwise operations as ((Pending | P.Force) and P.Mask) and then, select the highest prioritized interrupt. A high (low) bit in the Level register defines a high (low) priority for the corresponding interrupt line. On the same priority level, interrupt with larger line number is higher prioritized.

The handling of incoming interrupts depends on the Broadcast register. A high bit in Broadcast means that the corresponding interrupt should be broadcasted. A broadcasted interrupt is written to the Force register of all processors; therefore, it also has to be acknowledged by every processor. Otherwise an incoming interrupt is written to the shared Pending register. In this case, it is sufficient when one processor acknowledges the interrupt. A processor P can acknowledge an interrupt by writing to the *irq_ack* wire. The interrupt will be cleared from Pending as well as from P.Force. A bit in Pending can also be explicitly cleared by raising the corresponding bit in Clear.

Extended Interrupts IRQMP has only 15 outgoing interrupt lines (*irq_req*) for every processor. Therefore, all extended interrupts are mapped onto a single regular interrupt, determined by the value of $EIRQ \in \{1..15\}$ (with $EIRQ = 0$ indicating regular mode). Masking and prioritization of incoming extended interrupts are handled similarly to regular interrupts. When the interrupt $EIRQ$ is acknowledged by process P, in addition to clearing the Pending register, the line number of the extended interrupt is written into P.Extir, which allows to fetch the last acknowledged extended interrupt.

4.2.3.2 Formal Verification

For formal verification, we employ a reimplementation of the path-based symbolic simulation approach for SystemC proposed in [134] supporting XIVL. Due to the absence of a TLM property language (and a tool for property instrumentation), we manually build a formal verification environment. It consists of a symbolic input driver to IRQMP and a checker to monitor its responses, respectively. IRQMP is treated as black box during the verification. Our evaluation results consist of two parts.

Tackling Path Explosion in Symbolic Simulation Using a symbolic driver to make all inputs symbolic allows to explore all possible behaviors of IRQMP. However, in preliminary experiments, we ran into the path explosion problem in path-based symbolic simulation, as the number of feasible exploration paths grows exponentially. To mitigate this problem, we have implemented a path merging algorithm. Generally, merging results in more complex solver queries, while explicit exploration suffers from path explosion problem. Our merging strategy is based on manual annotations of branches and loops, which allow a fine grained selection between path merging and explicit path exploration. Additionally, we restricted the number of symbolic inputs based on the functionality under verification. We apply a number of different verification scenarios focusing on different aspects of IRQMP as described in the next section.

Verification Scenarios and Results Our symbolic verification scenarios are partitioned into three groups based on the functionality under verification: Broadcasting, Masking, and Prioritization of interrupts. Table 4.5 summarizes the verification results. The table shows the scenario ID, e.g. S1.1, along with a small description in the first column. The second column (*Verdict*) shows the verification result, where *Unsafe* indicates a bug in IRQMP and *Safe* indicates a successful verification. The last column shows the verification time in seconds. All experiments have been performed on an AMD 3.4 GHz machine running Linux. The time and memory limit has been set to 2 h and 6 GB, respectively.

The first group checks broadcast functionality. It contains a single scenario that verifies that IRQMP sends a broadcasted interrupt to every processor. For

Table 4.5 Verification results

Verification scenarios		Verdict	Time (sec.)
S1	Broadcast		
	S1.1 Sent to all CPUs	Safe	56.67
S2	Masking		
	S2.1 Single CPU with force	Unsafe	15.01
	S2.2 Two CPUs no force	Safe	24.15
	S2.3 Extended interrupts	Unsafe	24.84
S3	Prioritization		
	S3.1 Single request	Safe	13.79
	S3.2 With extended interrupts	Safe	17.09
	S3.3 Acknowledgement roundtrip	Safe	2093.42
	S3.4 With mask	Safe	3766.86
	S3.5 With mask and two CPUs	–	T.O.
	S3.6 With extended interrupts	Unsafe	306.48

this scenario, all 16 processors are used. The symbolic input driver selects non-deterministically the interrupt to be broadcasted (by setting the same bit in *irq_in* and Broadcast).

The Masking group verifies that every observed interrupt request for a processor P is not masked for P. The symbolic input driver in all S2 scenarios makes all Mask registers and *irq_in* symbolic. S2.1 fails because the Force register is erroneously applied after the Mask register during interrupt request computation in IRQMP. S2.3 uses two processors and extended interrupts (i.e. $EIRQ$ is additionally symbolic). It has revealed a subtle error that requires at least two processors: an $EIRQ$ interrupt can be sent to a processor even though all extended interrupts are masked. Non-formal verification techniques could easily miss this bug.

The Prioritization group checks whether interrupt requests come in order according to their priorities for every processor. In all scenarios of this group, the Level register and *irq_in* are set to be symbolic. The scenarios are constructed in increasing complexity order. S3.1 uses only one processor and no requested interrupts are acknowledged. S3.2 is similar but allows extended interrupts additionally. In the remaining four scenarios, all interrupts are acknowledged via either the *irq_ack* wires or the Clear register. Furthermore, S3.4 adds a symbolic Mask, S3.5 a second processor, and S3.6 extended interrupts. S3.6 detects an error in handling extended interrupts: extended interrupts are not automatically cleared when the $EIRQ$ interrupt is acknowledged; therefore, IRQMP keeps requesting extended interrupts without terminating.

4.2.3.3 Discussion and Future Work

We were able to prove important functional properties of IRQMP as well as to detect three major bugs. This clearly demonstrates the feasibility of formal verification of TLM peripherals. Conversely, the case study unveils open challenges to be addressed in future work, including, e.g.

- Incorporating a SystemC TLM property language into the formal verification flow to enable the automatic generation of symbolic input driver and checker;
- Additional language features to further close the gap between SystemC TLM and the current XIVL: similar to the added OO features this would allow to translate a larger set of SystemC programs without complex error-prone manual transformations;
- An intelligent path merging algorithm for symbolic simulation, which can automatically find a good balance between merging and explicit exploration, to replace the current one, which requires manual annotations.

4.3 Compiled Symbolic Simulation

Symbolic simulation which essentially combines symbolic execution [122] with *Partial Order Reduction* (POR) [57, 62] has been shown very effective in verifying SystemC designs. This combination enables complete exploration of the state space of a *Design Under Verification* (DUV), which consists of all possible values of its inputs and all possible schedules (i.e. orders of execution) of its (typically) asynchronous SystemC processes.

The existing symbolic simulators are commonly interpreter-based, i.e. they interpret the behavior of the DUV statement by statement. An interpreter-based symbolic simulator manages directly the execution state (i.e. the DUV state and the scheduler state) and updates it according to the semantics of the to-be-interpreted statement. The main advantage of this approach is the ease of implementation. Especially, when a conditional branch is encountered or multiple orders of execution of runnable processes are possible during the symbolic simulation, cloning the current execution state for further exploration is required and can be straightforwardly implemented. The trade-off for this convenience is the relatively low performance of the overall verification, which becomes clearer when symbolic simulation is applied to large models, e.g. the bare bone of a VP.

This book proposes *Compiled Symbolic Simulation* (CSS) to tackle this performance issue and makes the following contributions:

1. We describe a source code instrumentation technique—the main novelty of CSS—that augments the DUV to integrate the symbolic execution engine and the POR based scheduler. The augmented DUV can then be compiled using any standard C++ compiler, leveraging its efficient code optimizations, to create native binary code. This binary, when executed, can achieve the complete state space exploration much faster than interpreter-based techniques.
2. We incorporate a set of optimization techniques (e.g. path merging) tailored for CSS to improve the performance further.

An extensive set of experiments including comparisons with existing approaches demonstrates the benefits of CSS. We will use the XIVL example program shown in Fig. 4.16 as the running example to illustrate the CSS instrumentation.

To the best of our knowledge, SPIN [109] is the only verification tool that currently also supports compiled verification. SPIN translates a model in its input language Promela into an executable C program. Translation from SystemC to Promela has also been considered [28, 207]. In contrast to our symbolic approach, the program generated by these approaches, when compiled and executed, performs explicit model checking on the original Promela model.

```
1  int n;                                12    end_loop:
2  int sum;                              13  }
3                                        14
4  thread A {                            15  main {
5    int i = 0;                          16    sum = 0;
6    loop:                               17    n = ?(int);
7    if (i >= n) goto end_loop;          18    assume (n <= 9);
8      i += 1;                           19    start;
9      sum = sum + i;                    20    assert (sum <= 45);
10     wait_time(1);                     21  }
11     goto loop;
```

Fig. 4.16 Simple XIVL example program with a single thread that iteratively computes the sum for a given symbolic input value n. This will serve as the running example in the remainder of this section

4.3.1 *Overview*

An overview of the proposed approach is shown in Fig. 4.17. The starting point is an XIVL program, which is an intermediate representation of the SystemC DUV. Please note that the to-be-verified properties are already embedded into this program by means of assertions. Our approach first automatically translates the XIVL description into an executable C++ program, then invokes a C++ compiler to produce a native binary. Executing the binary equates to verifying the properties on the XIVL model with symbolic simulation. To accomplish this, several symbolic simulation components must be integrated into the XIVL to C++ translation. In the following, we discuss the main parts of the generated C++ program, which is also shown on the right-hand side of Fig. 4.17.

4.3.1.1 Generated C++ Program Overview

Essentially such a C++ program consists of four parts: (1) The scheduler; (2) The instrumented XIVL program; (3) The SMT layer; (4) The declaration of data structures to store and manipulate execution states. The *symbolic execution engine* (SEE) is located inside the instrumented program. This interacts with the scheduler, the SMT layer, and execution states via API functions to perform the state space exploration.

At each point of time during the execution, only one execution state is considered active. The scheduler provides API functions for managing a set of execution states including cloning (implemented as deep-copy) and swapping the currently active execution state. This allows to explore different independent paths, which arise due to process scheduling decisions and forking in symbolic branches. The scheduler selects which runnable thread or which direction of a branch to be executed next. It will also backtrack when the execution of a path finishes and thus eventually explore all different alternatives. Furthermore, the (static) dependency relation is computed

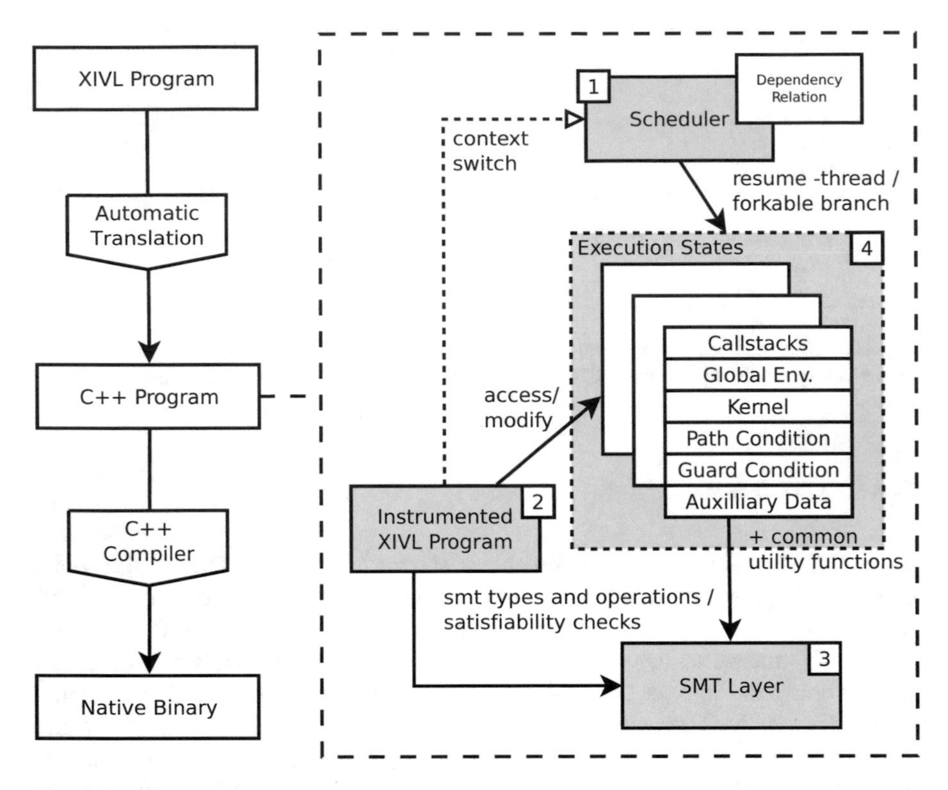

Fig. 4.17 CSS overview

during the translation process and embedded into the scheduler to prune redundant schedules.

The SMT layer provides API functions to create symbolic data types, to manipulate symbolic expressions, and to check their satisfiability. Essentially, it is a convenience layer on top of underlying SMT solvers.

The *main* function in the C++ program will first create the scheduler and then set up an initial execution state. Essentially this step registers all threads and the XIVL *main* function so that they can be selected by the scheduler to execute. In the following we will discuss the data structures for execution state and the instrumented XIVL code together with its interactions with other components in more detail.

4.3.1.2 Data Structures for Execution State

The first component of an execution state is the state of the simulation kernel (i.e. status of threads and events). The second component is the state of the XIVL model itself. This is further divided into global and local program state.

```
1  // callframes allow to preserve local execution state across context switches
2  struct Callframe {
3    unsigned ip;
4    void *env;
5    Callable fn;
6    void *result;
7  };
8  struct GlobalEnv { // global variables
9    SmtExpr n;
10   SmtExpr sum;
11 };
12 struct LocalEnv_A { // local variables of thread A
13   SmtExpr i;
14 };
15 // specialized callframe for every thread/function
16 struct Callframe_A : public Callframe {
17   LocalEnv_A env_A;
18   Callframe_f(Callable fn) {
19     env = &env_A;
20     result = nullptr; // threads have no return value
21     ip = 0;
22     fn = f;
23   }
24 };
```

Fig. 4.18 Data structures for the CSS transformation of thread A to store and access local and global program state (located in block 2 of Fig. 4.17)

All global variables are stored in a *GlobalEnv* class. In Fig. 4.18, Lines 8–11 show the definition of this class for the XIVL example. The local program state consists of the local state of all functions declared in the XIVL description. Please note that the local state of a function must be saved when a blocking statement (i.e. *wait*) is executed. This local state is stored in a callframe to preserve it across context switches. A callframe contains the following data: (a) the label identifier to resume execution at the interrupted point (IP); (b) a local environment (*env*), which stores arguments and local variables; (c) a pointer to a callable object, which is either a global function or a pair of class instance and member function, which allows to call the actual function; (d) a generic pointer to the result stored in the local environment, so a callee can retrieve it.

For every function a custom local environment and callframe class are statically generated during translation. The definition of the specific callframe class for the thread function A from the example is shown in Lines 16–24 in Fig. 4.18.

4.3.1.3 Instrumented XIVL Code

Essentially, the original XIVL code is instrumented in three steps: (1) Adding context switch logic to interact with the scheduler; (2) Replacing native data types and operations with symbolic types and operations; (3) Transforming branches to

```
1  #define GLOBAL(VAR) active_state->global_env->VAR
2  bool dir;
3  void thread_A(Callframe *c) {
4    // retrieve local state and goto beginning or last context switch location
5    LocalEnv_A *env = (LocalEnv_A *)c->local_env;
6    switch (c->ip) {
7      case 0: goto A_start;
8      case 1: goto A_branch_1;
9      case 2: goto A_interrupt_1;
10     default: assert (false);
11   }
12   A_start:
13   // create a constant SMT expression (compatible with symbolic SMT expressions)
14   env->i = smt_create(0);
15   loop:
16   // store location in case of context switch
17   c->ip = 1;
18   A_branch_1:
19   dir = active_state->on_branch(smt_ge(env->i, GLOBAL(n))); // check feasibility,
         might fork
20   if (dir) goto end_loop;
21     env->i = smt_add(env->i, smt_create(1));
22     GLOBAL(sum) = smt_add(GLOBAL(sum), env->i);
23     c->ip = 2;
24     wait_time (1); // modifies kernel state and triggers context switch
25     A_interrupt_1;
26     goto loop;
27   end_loop;
28   A_end:
29   c->ip = -1;
30 }
```

Fig. 4.19 CSS transformation of thread A (located in block 2 of Fig. 4.17), which has been defined in Fig. 4.16

support the exploration of both paths in case of a symbolic branch condition. The last two points are parts of the integrated SEE. In the following we will describe context switch, symbolic branch, and function call execution in more detail.

Context Switch Logic For each defined function, a label is associated with every statement that can interrupt the execution of the function. Every label is assigned a unique identifier within the function. A switch block is added at the beginning of the function to dispatch execution to the correct label. This block for our thread A can be seen in Lines 6–11 of Fig. 4.19. Before the dispatching, the local state of the function is retrieved (Line 5).

The corresponding label is added right after the blocking statement. An assignment of the IP of the callframe to the ID of this label is added right before the blocking statement. This allows to resume it after the context switch. These steps can be seen in Lines 23–25.

```
1  enum BranchDirection {True, False, None};
2  enum BranchStatus {BothFeasible, TrueOnly, FalseOnly};
3
4  bool ExecutionState::on_branch(const SmtExpr::pointer &cond) {
5    // initially *branch_direction* is set to None
6    if (branch_direction == BranchDirection::True) {
7      // remove scheduler annotation, update path condition accordingly and return
           concrete scheduler decision
8      branch_direction = BranchDirection::None;
9      PC = smt->bool_and(PC, cond);
10     return true;
11   } else if (branch_direction == BranchDirection::False) {
12     // similar to above branch direction
13     branch_direction = BranchDirection::None;
14     PC = smt->bool_and(PC, smt->bool_not(cond));
15     return false;
16   }
17   assert (branch_direction == BranchDirection::None);
18
19   // let the scheduler decide in case both branches are feasible
20   BranchStatus::e b = check_branch_status(cond);
21   if (b == BranchStatus::BothFeasible) {
22     forkable = true;
23     // back to the scheduler, unwind potentially nested function
24     throw ContextSwitch(active_state);
25   } else {
26     return b == BranchStatus::TrueOnly ? true : false;
27   }
28 }
```

Fig. 4.20 Execution of branches with symbolic conditions (located in block 4 of Fig. 4.17)

Executing Symbolic Branches The branch in Line 7 of Fig. 4.16 is translated into the code block shown in Lines 17–20 of Fig. 4.19. Please note that at this point, native data types and operations have been already replaced with symbolic types and operations (prefixed with *smt* in Fig. 4.19). The *on_branch* function of the active execution state is called with the symbolic condition and will always return a concrete Boolean value. This result is stored in a designated global variable *dir* which is used as concrete condition in the branch.

The *on_branch* function is shown in Fig. 4.20 and works as follows: First the SMT layer is invoked to check if both paths are feasible in Line 20. If only one path is feasible, then the corresponding concrete Boolean value—true for the goto path and false for the fallthrough path—is returned in Line 26 and the execution will continue normally without any interruption. In case both paths are feasible, the execution state will be marked as forkable and a context switch back to the scheduler will be triggered (Lines 22–24). The scheduler will then clone the execution state and decide which path to continue. The decision will be annotated on the execution state and execution is resumed by calling the interrupted thread function of A. The execution of thread A will directly continue at the branch statement since the IP of A has been updated to point to the branch label in Line 18 during the first execution.

The *on_branch* function is called again, but this time the execution state is marked with a concrete path decision annotated by the scheduler, i.e. either the condition in Line 6 or Line 11 will be true. In this case the function will ignore the symbolic condition, but instead simply return the scheduler decision and update the path condition of the active execution state accordingly. Additionally, it will remove the scheduler annotation from the state to ensure the correct handling of the next branch.

Function Calls Every function is translated into two parts: a wrapper and an implementation. The original function call is replaced by a call to the wrapper function which in turn calls the implementation. An example for the function call $x = f(y,z) = y + z$ is shown in Fig. 4.21. The wrapper (Line 8) has the same signature, except the return type, as the original function. It allocates a callframe based on the given arguments and delegates to the implementation function (Line 12 and Line 1). A wrapper is used because the target function is not statically known when calling a virtual function.

The implementation function contains the logic of the original function, instrumented similarly as the example for thread A has demonstrated (see Fig. 4.19). Once the implementation function finishes, execution will continue at the callee in Line 18. The callee can retrieve the result value from the callframe and is responsible for cleanup (Lines 20–22). A label is placed after each function call (Line 18) and the IP is set to that label right before the function call (Line 16) to resume execution correctly in case the implementation function is interrupted. In this case execution

```
1  void f_impl(Callframe *c) {
2    LocalEnv_f *env = (LocalEnv_f *)(c->local_env);
3    // ... other instructions
4    env->result = smt_add(env->a, env->b);
5    goto f_end;
6    // ... other instructions
7  }
8  void f_wrapper(SmtExpr a, SmtExpr b) {
9    // wrapper allows correct callframe allocation in case of virtual functions
10   Callframe_f *c = new Callframe_f(a, b, f_impl);
11   active_state->push_callframe(c);
12   f_impl(c);
13 }
14
15 // implementation function does not return here when interrupted, thus store
       resumption point
16 c->ip = LABEL_ID(f_call_1);
17 f_wrapper(env->y, env->z);
18 f_call_1:
19 // retrieve the result and cleanup
20 Callframe *c_f = active_state->pop_callframe();
21 env->x = *((SmtExpr *)(c_f->result));
22 delete c_f;
```

Fig. 4.21 Relevant instructions for a function call (located in block 2 of Fig. 4.17)

control will return to the scheduler and unwind the native C++ stack. The scheduler is then responsible to call the callee once the implementation function finishes.

4.3.2 *Optimizations*

This section presents two optimizations tailored for CSS. First, we show how to efficiently integrate existing path merging methods into our CSS framework. This integration is important, since path merging is a powerful technique to alleviate the path explosion problem during symbolic execution. Second, we show how to generate more efficient code by employing a static analysis to determine which operations can be executed natively. This can further speed up the execution of concrete code parts significantly.

4.3.2.1 Path Merging

This section describes how to use path merging in combination with CSS. Our algorithm assumes that branches and loops that should be merged are marked, e.g. by placing *@mergeable* before them in the input code. This is a flexible approach that allows both an automatic analysis, e.g. see [129], and a user to annotate mergeable branches. This allows a fine grained tuning between merging and explicit exploration. Merging can reduce the number of explored paths exponentially at the cost of more complex solver queries. A limitation of our current algorithm is that merging SystemC kernel parts is not yet supported, since the kernel state is available in explicit form. Therefore, loops and branches that update kernel state will not be merged. In the following we will discuss our branch merging approach in more detail. Loop merging in principle works similar to branch merging and thus will not be further discussed.

Merging Branches Consider an example program *x=?(int); if (x > 0) { x = 2*x; } else { x = 1; } x++;*, where the *if-statement* is marked as mergeable. This code is transformed into the code block shown in Fig. 4.22. The initial assignment of x and the increment at the end are not inside the mergeable block and thus normally translated.

Four functions are involved in the translation of mergeable branches, for convenience we use short names in this example: *begin*, *resume*, *end*, and *assign*. They operate on the currently active execution state. The additional global variable *resume_merging* is a simple Boolean value, similar to dir, to control execution of the program. It allows to distinguish between the first and second visit of the branch. Initially it is set to *false*. The execution state internally keeps a stack of triples *(GC, status* $\in$ *{none, first, second}, cond)* to capture the execution progress of potentially nested mergeable branches. GC is the current guard condition and *cond* the symbolic branch condition passed to the *begin* function. Initially the guard

```
1  env->x = smt_bv(32, true);
2  START_LABEL:
3  if (resume_merging) {
4    resume_merging = false;
5    dir = resume();
6  } else {
7    dir = begin(smt_gt(env->x, smt_create(0)));
8  }
9  if (dir) {
10   env->x = assign(env->x, smt_mul(env->x, smt_create(2)));
11 } else {
12   env->x = assign(env->x, smt_create(1));
13 }
14 if (end()) {
15   resume_merging = true;
16   goto START_LABEL;
17 } else {
18   resume_merging = false;
19 }
20 env->x = smt_add(env->x, smt_create(1));
```

Fig. 4.22 Branch merging example (located in block 2 of Fig. 4.17)

condition is *true*. On every solver query, it is combined with the path condition using a logic *and* operation. There are two choices when *begin* is called:

(1) Both paths are feasible. The begin function will store the triple *(GC, first, cond)* and return *true*, an arbitrary choice to start execution with the *if-path*. Backing up the guard condition and branch condition allows to use them for the exploration of the *else-path*, as they can be modified during execution of the *if-path*. Furthermore, *GC* is updated as $GC \wedge cond$. The *assign(lhs, rhs)* function will return a guarded expression of the form *GC ? rhs : lhs*, i.e. update lhs to rhs based on GC. Finally the *end* function will notice that *status=first*, i.e. branch merging is active, so it will update status to *second* and return *true*. Therefore, execution will jump to the beginning of the branch again, but this time the *resume* function will be entered. It will retrieve the stored *GC* and *cond* from the top of stack, update *GC* as $GC \wedge \neg cond$ and return *false* to execute the *else-path*. Since *status=second*, the *end* function will pop the top of stack and return false, which leaves the mergeable branch.
(2) Only one path is feasible. The begin function will push *(GC, none, cond)* on the stack and return the path direction. The *assign* function will either return *GC ? rhs : lhs*, in case this branch is nested within another active mergeable branch, i.e. $GC \neq true$, or just rhs, otherwise. The end function will recognize *status=none*, thus pop the data triple and return *false*.

4.3.2.2 Native Execution

We provide two native execution optimizations to improve the execution performance of instructions and function calls, respectively: (1) A static analysis is employed to determine variables that will never hold a symbolic value. Such variables can keep their native C++ type. Furthermore, native operations can be performed on native data types. They are significantly faster than (unnecessarily) manipulating symbolic expressions. (2) To optimize function calls, a static analysis is employed to determine which functions can be interrupted. Essentially, a function is interruptible, iff it contains an interruptible statement, i.e. *wait_time/event* or branch with symbolic condition, or any function it calls is interruptible. Callframe allocation and cleanup are not required for a non-interruptible function; therefore, a function call $x = f(y, z)$ is not instrumented but executed unmodified on the native stack. In the following we describe our first static analysis, which determines symbolic variables, in more detail.

Computing Maybe-Symbolic Variables This static analysis starts by computing two pieces of information: (1) A root set of variables which are symbolic. (2) A dependency graph between variables, where an edge from a to b denotes that b maybe-symbolic if a maybe-symbolic. Then all variables reachable from the root set following the dependency graph can potentially be symbolic. Such variables are called *maybe-symbolic* The other variables can keep their native data types.

The root set S contains all variables where a symbolic value is directly assigned, e.g. $x =?(int)$ will add x to S. Similarly $*p =?(int)$ will add the pointer p to S, since it needs to have *SmtExpr** type.

The dependency graph G essentially records assignments between variables. Here pointer and non-pointer types are treated differently. The assignment $a = b$ will add an edge from b to a. Conversely, if a is symbolic, then b can still be a native type, since b can simply be wrapped in an *smt_create* call to be compatible with a. This allows native execution of other operations involving b. If a pointer is involved in the assignment, e.g. $p = \&a$, then an edge in each direction is added to G. The reason is that there is no conversion available that allows to assign a *int** to *SmtExpr**. Therefore a must also have a symbolic type. Argument and return value passing during function calls are handled the same way as assignments.

Determining Symbolic Operations Once the set of maybe-symbolic variables is known, it can be used to compute the set of symbolic expressions by walking the expression trees bottom-up. This allows to determine which operations need to be performed by the symbolic execution engine and which can be natively executed. For example, consider the expression $a < b + 1$, where a is symbolic and b is not. Then $b + 1$ will first be natively executed and then converted to a symbolic value. Finally the $<$ operation will be executed symbolically.

4.3.3 *Experiments*

We have implemented the proposed CSS together with the optimizations and evaluated it on an extensive set of benchmarks. The evaluation also includes comparisons to state-of-the-art tools. We employ Z3 v4.4.1 in the SMT Layer and compile the generated C++ programs using Clang 3.8 with $-O3$ optimization. All experiments were performed on an Intel 3.5 GHz machine running Linux. The time and memory limit has been set to 2000s and 6 GB, respectively. In the following tables all runtimes are given in seconds. T.O. (M.O.) denotes that the time (memory) limit has been exceeded. The column V (Verdict) denotes if the benchmark is bug-free, i.e. safe (S), or contains bugs (U). Thus, if a runtime can be reported for a tool on a benchmark with (without) bug, it means the tool terminated successfully and detected the bug in the benchmark (confirmed its correctness) as expected.

4.3.3.1 Native Execution Evaluation

Our first experiment demonstrates the benefits of native execution over interpretation in the context of symbolic execution. We compared CSS with KLEE [27], the state-of-the-art symbolic executor for C, which includes a highly optimized interpretation engine for the LLVM-IR. The results are shown in Table 4.6. For CSS, we show the compilation time (column *Compile*) and include it into the total runtime (column TOTAL) to make the comparison fair. The upper half of the table shows pure C benchmarks, for which KLEE is directly applicable. The *iterative* and *recursive* benchmarks perform some lightweight symbolic computation in 4 (*small*) and 400 (*large*) million iterations/recursive calls, respectively.

The lower half of Table 4.6 shows SystemC benchmarks. Since KLEE cannot directly handle SystemC semantics, we applied a sequentialization scheme similar to [38, 72]. For a fair comparison, we limited the state space to a single arbitrary

Table 4.6 Native execution evaluation (runtimes in seconds)

			CSS	
Benchmark	V	KLEE	TOTAL	Compile
Iterative-small	S	4.60	0.62	0.61
Iterative-large	S	444.64	0.84	0.62
Recursive-small	S	104.77	0.65	0.61
Recursive-large	S	M.O.	0.85	0.61
mem-slave-tlm-bug.500K*	U	15.16	0.81	0.74
mem-slave-tlm-bug.5M*	U	144.56	0.83	0.58
mem-slave-tlm-sym.500K*	S	17.17	10.91	0.71
mem-slave-tlm-sym.5M*	S	164.65	113.17	0.65
pressure.40M*	S	23.03	0.82	0.58
pressure.400M*	S	220.74	3.08	0.59

process schedule; otherwise, KLEE would perform very poorly because it does not support POR. Both KLEE and CSS are then applied on the same sequentialized programs available in C or XIVL, respectively. These sequentialized benchmarks are marked with *. For scalability investigation, we also varied the size of the benchmarks, indicated by the last number in benchmark name with *K=thousand* and *M=million*. In these cases, that means the number of loop iterations and the maximum simulation time.

Overall, CSS shows significant improvements over KLEE. The *mem-slave-tlm-sym* benchmark performs in every loop iteration heavier symbolic computations, which are not optimized by native execution. Therefore, the benefit of CSS is less pronounced.

4.3.3.2 Comparison with Existing SystemC Verifiers

Freely Available Benchmarks Table 4.7 shows a comparison between CSS with *Interpreted Symbolic Simulation* (ISS), basically a reimplementation of the symbolic simulation technique described in Sect. 4.1.2, and the state-of-the-art abstraction-based verifier KRATOS [38]. The benchmarks are freely available and

Table 4.7 Comparison with existing SystemC verifiers on publicly available benchmarks (runtimes in seconds)

Benchmark	V	ISS	KRATOS	CSS
jpeg-p6-bug	U	2.57	T.O.	1.29
mem-slave-tlm-bug.50	U	2.74	T.O.	0.84
mem-slave-tlm-bug.500K	U	M.O.	T.O.	19.96
mem-slave-tlm-bug.5M	U	M.O.	T.O.	208.81
mem-slave-tlm-sym.50	S	2.81	T.O.	0.88
mem-slave-tlm-sym.500K	S	M.O.	T.O.	47.96
mem-slave-tlm-sym.5M	S	M.O.	T.O.	479.70
pressure-safe.10	S	7.18	0.41	0.71
pressure-safe.15	S	211.36	1.27	1.42
pressure-safe.40M	S	M.O.	T.O.	M.O.
pressure-unsafe.25	U	0.83	31.79	0.67
pressure-unsafe.50	U	0.82	443.08	0.68
simple-fifo-bug-1c2p.20	U	1.33	18.10	0.92
simple-fifo-bug-1c2p.50	U	1.94	1806.62	0.85
simple-fifo-bug-2c1p.20	U	1.45	14.54	0.90
simple-fifo-bug-2c1p.50	U	2.30	434.98	0.84
token-ring-bug2.15	U	1.59	3.74	1.66
token-ring-bug2.20	U	1.92	M.O	1.90
token-ring-bug.20	U	1.07	149.26	1.19
token-ring-bug.100	U	4.20	M.O	3.93

commonly used to compare SystemC formal verification tools (see, e.g. [38, 134]). The comparison with KRATOS is mainly to confirm that our ISS implementation is reasonably fast. Generally, the obtained results are consistent with the results reported in [134]. Again, we also varied the size of the benchmarks for scalability investigation. The compilation times are already integrated into the CSS runtimes. In general they are negligible for longer running benchmarks. For CSS, improvements of several orders of magnitude can be observed. In one case CSS is unable to verify the up-scaled *pressure* benchmark, since our POR algorithm is unable to limit the exponential growth of thread interleavings due to complex dependencies. This problem can be solved by employing a stronger POR algorithm or combine CSS with a stateful exploration.

Real-World Virtual Prototype Models This second comparison was performed on two larger real-world SystemC VP models: (1) An extended version of the Y86 CPU [16] which implements a subset of the instructions of the IA-32 architecture [114]. (2) A TLM model of the *Interrupt Controller for Multiple Processors* (IRQMP) of the LEON3-based VP SoCRocket, partly developed and used by the European Space Agency [189].

Since both models extensively use object-oriented programming features as well as arrays, KRATOS is not applicable and thus omitted from the comparison. The results are shown in Table 4.8. For CSS, the first column shows the total runtime. The next three columns show the detailed breakdown (including percentage) of the total time into native execution time, SMT solving time, and compilation time. In an analogous manner, the ISS total runtime as well as its breakdown into interpretation time and SMT solving time is reported. Also please note that both CSS and ISS explore the state space in the same order, i.e. they follow the same execution paths and solve the same SMT queries.

Benchmarks in the upper half only operate on concrete values (hence the SMT time is not available). The *test-1*, *test-2*, and *test-3* are testcases for the IRQMP from the SoCRocket distribution. The *y86-counter* benchmarks execute a computation on the Y86 processor model. As expected, CSS significantly outperforms ISS here.

The lower half shows results for the verification of functional properties on the IRQMP model. The *y86-isr* benchmark combines the IRQMP and Y86 processor model. The Y86 model runs a version of the counter program and furthermore an *Interrupt Service Routine* (ISR) is placed in memory. The IRQMP prioritizes a symbolic interrupt received from the test driver and forwards it on a signal line to the Y86 CPU. The CPU stores the received interrupt in its register and triggers the ISR, which processes and acknowledges the interrupt using memory mapped IO over a bus transaction. The other benchmarks in the lower half verify functional properties of the IRQMP, in particular that broadcasts are sent to all CPUs and interrupt prioritization as well as masking works correctly.

For the benchmarks in the lower half, notable improvements in total runtimes can still be observed, although not in the scale of the previous comparison. The reason becomes clear when inspecting the detailed breakdown of runtimes. While significant improvements of native execution over interpretation are still visible,

Table 4.8 Real-World Virtual Prototype Benchmarks (runtimes in seconds)

		CSS							ISS				
Benchmark	V	TOTAL	Execution		SMT		Compilation		Total	Interpretation		SMT	
Test-1	S	4.49	0.50	11%	–		3.99	89%	79.28	79.28	100%	–	
Test-2	S	4.15	0.23	5%	–		3.93	95%	44.52	44.52	100%	–	
Test-3	S	3.88	0.24	6%	–		3.64	94%	35.99	35.99	100%	–	
y86-counter	S	0.94	0.12	13%	–		0.82	87%	276.70	276.70	100%	–	
Broadcast	S	10.54	0.62	6%	6.69	63%	3.23	31%	18.68	10.15	54%	8.53	46%
Prioritization-8	S	34.01	0.98	3%	30.22	89%	2.81	8%	114.20	70.58	62%	43.62	38%
Prioritization-12	S	497.36	16.96	3%	477.63	96%	2.77	1%	T.O.	N.A.	N.A.	N.A.	N.A.
Masking-regular	U	5.69	0.10	2%	2.49	44%	3.10	54%	7.22	4.28	59%	2.94	41%
Masking-extended	U	7.81	0.36	5%	4.22	54%	3.23	41%	10.69	5.48	51%	5.21	49%
y86-isr	S	179.83	30.79	17%	140.14	78%	8.90	5%	T.O.	N.A.	N.A.	N.A.	N.A.

the SMT solving times for CSS are only slightly better. Conversely, path merging has been shown to be crucial to avoid path explosion in verifying properties on the IRQMP design. Solving complex SMT queries, as a result of extensive path merging, often dominates the total runtime for these benchmarks. Thus, the overall advantage of CSS over ISS is less pronounced here.

Also please note that while these real-world SystemC VP models are still rather small compared to those that can be verified by simulation-based techniques, we are unaware of any other symbolic verification approach for SystemC, which can scale to models of this complexity (e.g. *y86-isr*).

4.3.4 Discussion and Future Work

Even though CSS and further optimizations such as path merging have remarkably improved the scalability of symbolic simulation for SystemC, it is still expectedly subject to state space explosion on large SystemC VP models. The two main limitations of our CSS approach are as follows: First, the native execution selection of CSS is currently based on a static (type) analysis. This prevents optimization in case code parts, e.g. functions, are executed multiple times with concrete and symbolic parameters. Second, CSS is currently stateless, i.e. it does not keep a record of already executed states, and cannot yet be used to verify safe programs with cyclic state spaces. It is still to be investigated, how a stateful symbolic simulation approach, such as described in Sect. 4.1, can be efficiently incorporated into CSS.

4.4 Parallelized Compiled Symbolic Simulation

This section presents the tool *ParCoSS* which is based on CSS and additionally supports parallelization to further improve simulation performance. Compared to the original CSS approach ParCoSS uses a *fork/join* based state space exploration instead of manually cloning the execution states to handle non-deterministic choices due to symbolic branches and scheduling decisions. A *fork/join* based architecture most notably has the following advantages: (1) It allows to generate more efficient code. (2) It drastically simplifies the implementation.

In particular, we avoid the layer of indirection necessary for variable access when manually tracking execution states and use native execution for all function calls by employing coroutines. Besides very efficient context switch implementation, coroutines allow natural implementation of algorithms without unwinding the native stack and without using state machines to resume execution on context switches. Additionally, manual state cloning of complex internal data structures is error prone and difficult to implement efficiently, whereas the *fork* system call is already very mature and highly optimized. Finally, our architecture allows for straightforward and efficient parallelization by leveraging the process scheduling and memory sharing capabilities of the underlying operating system.

4.4.1 Implementation Details

To simplify development, facilitate code reuse and the translation process from XIVL to C++ we have implemented the PCSS (Parallel CSS) library, which provides common building blocks for parallel symbolic simulation. The PCSS library is linked with the C++ program during compilation. An overview of our tool is shown in Fig. 4.23. In the following we will describe our PCSS library, provide more details on the *fork/join* based exploration, and briefly sketch the translation process from XIVL to C++.

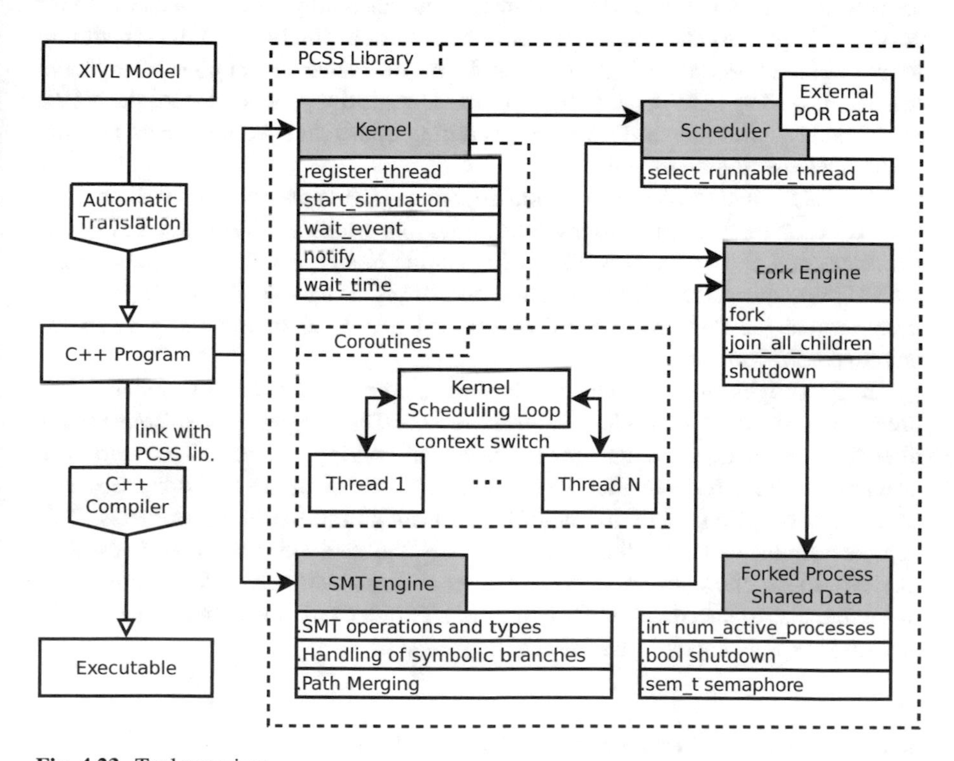

Fig. 4.23 Tool overview

4.4.1.1 PCSS Library

The right-hand side of Fig. 4.23 shows the main components, and their interaction, of the PCSS library. Essentially it consists of the following components: kernel, scheduler, SMT engine, fork engine, and some process shared data.

The kernel provides a small set of functions which directly correspond to the XIVL kernel related primitives (e.g. *wait* and *notify*) and allows to simulate the SystemC event-driven simulation semantics. Furthermore, all thread functions of the XIVL model are registered in the kernel. The kernel will allocate a coroutine with every thread function as entry point. Coroutines naturally implement context switches as they allow to jump execution between arbitrarily nested functions while preserving the local data. Our implementation is using the lightweight *boost context* library and in particular the functions *make_fcontext* and *jump_fcontext* to create and switch execution between coroutines, respectively. The scheduler is responsible for selecting the next runnable thread inside the scheduling loop of the kernel. Our coroutine implementation allows to easily switch execution between the scheduling loop and the chosen thread. POR is employed to reduce the number of explored interleavings. The POR dependency relation is statically generated from the XIVL model and encoded into the C++ program during translation. At runtime it is passed to the scheduler during initialization.

The SMT engine provides common functionality required for symbolic execution. It keeps track of the current path condition, handles the *assume* and *assert* functions, and checks the feasibility of symbolic branch conditions. Furthermore, the SMT engine provides SMT types and operations. Essentially this is a lightweight layer around the underlying SMT solver and allows to transparently swap the employed SMT solver.

The fork engine is responsible to split the execution process into two independent processes in case of a non-deterministic choice. This happens when both branch directions are feasible in the SMT engine or multiple thread choices are still available in the scheduler after applying POR. One of the forked processes will continue exploration, while the other is suspended until the first terminates. This approach simulates a depth first search (DFS) of the state space. As an optimization, the fork engine allows to run up to N processes in parallel, where N is a command line parameter to the compiled C++ program. Parallelization is very efficient as the processes explore disjoint state spaces independently.

4.4.1.2 Fork/Join Based State Space Exploration

Executing Symbolic Branches The *on_branch* function in the SMT engine, shown in Fig. 4.24, accepts a symbolic branch condition and returns a concrete decision, which is then used to control native C++ control flow structures. The *check_branch_status* checks the feasibility of both branch directions by checking

```
1  bool on_branch(const SmtExpr &cond) {
2    auto stat = check_branch_status(cond);
3    if (stat == BranchStatus::BothFeasible) {
4      bool is_child = fork_engine->fork();
5      if (is_child)
6        pc = smt->bool_and(pc, cond);
7      else
8        pc = smt->bool_and(pc, smt->bool_not(cond));
9      return is_child;
10   }
11   return stat == BranchStatus::FalseOnly ? false : true;
12 }
```

Fig. 4.24 Symbolic branch execution

```
1  bool ForkEngine::fork() {
2    int pid = ::fork();
3    if (pid != 0) {
4      num_children++;
5      while (!try_fork(shared_data, N)) {
6        if (num_children > 0) {
7          join_any_child();
8        } else {
9          usleep(1); // wait for someone else to join child
10       }
11     }
12   } else {
13     num_children = 0;
14   }
15   return pid == 0;
16 }
```

Fig. 4.25 Implementation of parallelized forking

the satisfiability of the branch condition and its negation. In case both branch directions are feasible, the execution will fork (Line 4) into two independent processes and update the path condition (pc) with the normal (Line 6) or negated condition (Line 8), respectively. Please note that the execution is not forked and the path condition is not extended when only a single branch direction is feasible.

Parallelization The forked processes communicate using anonymous shared memory, which is created during initialization in the first process using the *mmap* system call and thus accessible from all forked child processes. The shared memory essentially contains three information: (1) counter variable to ensure that no more than N processes will run in parallel, (2) shutdown flag to gracefully stop the simulation, e.g. when an assertion violation is detected. (3) unnamed semaphore to synchronize access. The semaphore is initialized and modified using the *sem_init*, *sem_post*, and *sem_wait* functions. Furthermore, each process locally keeps track of the number of forked child processes (*num_children*). Figure 4.25 shows an implementation of the *fork* function. First the *fork* system call is executed. The child process (*pid* is zero) will never block, since executing only one process will not increase the number of active processes. The parent process however will first try to atomically check and increment the shared counter in Line 5. When this fails, i.e. the maximum number N of processes is already working, the parent process will wait until a working processes finishes, by either awaiting one of its own children (Line 7) or until some other process joins its children (Line 9).

4.4.2 Evaluation and Conclusion

We have evaluated our tool on a set of SystemC benchmarks from the literature [20, 38, 134] and the TLM model of the *Interrupt Controller for Multiple Processors* (IRQMP) of the LEON3-based virtual prototype SoCRocket [189]. All experiments have been performed on a Linux system using a 3.5 GHz Intel E3 Quadcore with Hyper-Threading. We used Clang 3.5 for compilation of the C++ programs and Z3 v4.4.1 in the SMT Layer. The time (memory) limit has been set to 750 s (6 GB), respectively. T.O. denotes that time limit has been exceeded. The results are shown in Table 4.9. It shows the simulation time in seconds for Kratos [38], *Interpreted Symbolic Simulation* (ISS) [36, 134] (also described in Sect. 4.1.2), and our tool ParCoSS with a single process (P-1) and parallelized with four (P-4) and eight processes (P-8). Comparing P-4 with P-8 allows to observe the effect of Hyper-Threading. The results demonstrate the potential of our tool and show that our parallelization approach can further improve results. As expected, CSS can be considerably faster than ISS. On some benchmarks Kratos is faster due to its abstraction technique, but the difference is not significant. Furthermore, Kratos is not applicable to the *irqmp* benchmark due to missing C++ language features. For future work we plan to integrate dynamic information, for POR and selection of code blocks for native execution, into our CSS framework.

Table 4.9 Experiment results, T.O. denotes timeout (limit 750s)

			ParCoSS		
Benchmark	Kratos	ISS	P-1	P-4	P-8
buffer-ws-p5	1.400	65.951	9.086	2.882	1.987
mem-slave-tlm-bug-50	T.O.	3.731	<0.1	<0.1	<0.1
mem-slave-tlm-sym-50	T.O.	3.940	<0.1	<0.1	<0.1
pressure-15	1.281	219.300	17.182	5.312	3.855
pressure-bug-50	444.781	0.897	<0.1	<0.1	<0.1
irqmp-8	–	108.670	32.719	10.815	8.237
irqmp-12	–	T.O.	530.705	178.108	128.257

4.5 Summary

This chapter presented approaches based on symbolic simulation to improve the scalability and applicability of formal verification for SystemC VPs.

First we presented a *stateful symbolic simulation* approach for the efficient and complete verification of safety properties in high-level SystemC designs. We expect that the SystemC design is available as an IVL description. The IVL is compact and easily manageable but at the same time powerful enough to model the behavior of SystemC designs. With the presented verification approach, safety properties can be completely proven on cyclic state spaces, which is not possible with a stateless symbolic simulation. The well-known state explosion problem is alleviated by integrating two complementary reduction techniques with SymEx, namely POR and SSR. The former allows to prune redundant schedules, whereas the latter can increase the effectiveness of symbolic state matching significantly. In addition to an exact algorithm for subsumption detection called ESS, we present several optimizations to improve its efficiency. We introduced a novel cycle proviso specifically tailored for SSR to ensure soundness in combination with POR. The experiments using an extensive set of benchmarks demonstrated the efficiency of these optimizations and the competitiveness of the stateful symbolic simulation with the state of the art.

In the next step we presented a case study on formal verification of TLM peripheral models using our symbolic simulation techniques. First, we showed how to bridge the gap between the industry-accepted modeling pattern for TLM peripheral models and the semantics of formal verification approaches for SystemC. Then, we reported verification results for the interrupt controller of the LEON3-based SoCRocket VP. We were able to prove important functional properties of IRQMP as well as to detect three major bugs which clearly demonstrates the feasibility of formal verification of TLM peripherals.

To further boost the scalability of symbolic simulation for SystemC verification we proposed *Compiled Symbolic Simulation* (CSS). In contrast to existing symbolic simulation approaches CSS is based on compiled execution instead of interpretation. For a scalable exploration, the symbolic execution engine and the *Partial Order Reduction* (POR) based scheduler are integrated into the DUV. Then a standard C++ compiler is used to generate a native binary, which will perform an efficient exhaustive exploration of the DUV. To further improve the efficiency we have proposed two optimizations tailored for CSS: existing path merging methods adapted to CSS in order to mitigate the well-known state explosion problem in symbolic simulation, and native execution, which can further speed up the execution of concrete code parts significantly. The experiments using an extensive set of freely available SystemC benchmarks as well as larger real-world SystemC TLM models demonstrated the efficiency and applicability of our approach. Finally, we integrated additional optimizations based on parallelization and architectural improvements into CSS and demonstrated that parallelization can achieve further significant performance improvements.

Chapter 5
Coverage-Guided Testing for Scalable Virtual Prototype Verification

Despite the recent progress in formal verification of SystemC designs as presented in the previous chapter, simulation-based verification is still the method of choice for SystemC-based VPs in industrial practice thanks to its scalability and ease of use. Basically, a set of stimuli is applied to the *Design Under Verification* (DUV; which can be either a whole VP, a set of components, or a single component) and for each stimulus, the actual behavior is checked against the expected behavior (e.g. specified by reference outputs or temporal properties). Since a VP is in essence a software model, simulation-based verification for VPs is actually very similar to software testing and therefore, techniques from the software domain can be borrowed to ensure a high quality of verification results. For example, high statement coverage of the DUV implied by the set of stimuli (also referred to as *testsuite* in the remainder of the chapter to reflect the software testing point of view) is nowadays a minimum requirement. This chapter presents two approaches for VP verification using coverage-guided test-case generation. Both approaches have been shown very successful in the SW domain to generate a comprehensive testsuite with a high error detection rate.

We start in Sect. 5.1 by presenting *Data Flow Testing* (DFT) for SystemC-based VPs. The contribution is twofold: First, we develop a set of SystemC specific coverage criteria for DFT. This requires to consider the SystemC semantics of using non-preemptive thread scheduling with shared memory communication and event-based synchronization. Second, we explain how to automatically compute the data flow coverage result for a given VP using a combination of static and dynamic analysis techniques. The coverage result provides clear suggestions for the testing engineer to add new testcases in order to improve the coverage result. The experimental results on real-world VPs demonstrate the applicability and efficacy of the analysis approach and the SystemC specific coverage criteria to improve the testsuite. The approach has been published in [79].

Then, Sect. 5.2 presents a novel *Coverage-Guided Fuzzing* (CGF) approach to improve the test-case generation process for verification of *Instruction Set*

V. Herdt et al., *Enhanced Virtual Prototyping*,
https://doi.org/10.1007/978-3-030-54828-5_5

Simulators (ISS) which is a crucial component in every VP. In addition to code coverage we integrate functional coverage and a custom mutation procedure tailored for ISS verification. Fuzzing is particularly useful to trigger and check for error-cases and can complement other test-case generation techniques. As a case study we apply our approach on a set of three publicly available RISC-V ISSs. We found several errors, including one error in the official RISC-V reference simulator *Spike*. This approach has been published in [102].

5.1 Data Flow Testing for Virtual Prototypes

Although a necessary step, statement coverage (and also stronger code coverage metrics) have some well-known limitations in their capability to detect bugs as well as to reflect the thoroughness of verification. In the software testing community, a fault-based technique with better bug detection capability, known as mutation analysis [43, 78], has been considered for decades. Essentially, mutation analysis measures the adequacy of the testsuite with respect to its ability to detect a set of injected faults, which are introduced to the program under test by applying small syntactic changes. The key to effectiveness is a fault model that is both simple and representative for typical coding mistakes. The ideas have also been successfully transferred to hardware verification as well as to SystemC. For instance, dedicated fault models for SystemC-based mutation analysis have been proposed in [22, 23, 190]. Mutation based solutions for software-driven verification have also been presented [73]. Commercial mutation analysis tools with support for SystemC such as Certitude from Synopsys are also available.

Considering the successful adoption of mutation analysis, it is rather surprising that another effective testing technique known as *Data Flow Testing* (DFT) [130, 175], to the best of our knowledge, has not been yet considered for SystemC. DFT also holds the promise of better bug detection capability. The underlying idea of DFT is that the propagation of (wrong) data is a necessity to reveal bugs: If a line of code produces a wrong value, the execution after that point must include another line of code that uses this erroneous value, otherwise there will be no observable failures. Based on this information, researchers have proposed several data flow adequacy criteria. These criteria require the testsuite to sufficiently exercise the identified *definition-use pairs*, i.e. pairs of definition (a statement where a value is produced) and use (a statement where this value is used). Recent research in DFT focused on automated test generation for data flow adequacy [201, 213] and on extending DFT to object-oriented programs [7, 45].

In this section we present a DFT approach for SystemC-based VPs, which does not come without challenges. A SystemC DUV is essentially a concurrent program with non-preemptive thread scheduling, shared memory communication, and event-based synchronization. This unique combination requires rethinking of the known DFT techniques. To this end, our contribution is twofold: First, we develop a set of SystemC specific coverage criteria for DFT, which takes the non-preemptive context switches and synchronization primitives of SystemC into consideration. Second, we explain how to automatically compute the data flow coverage result for a DUV using a combination of static and dynamic analysis. The coverage result provides clear suggestions for the testing engineer to add new testcases in order to improve the coverage result. Our experimental results on real-world VPs demonstrate the applicability and efficacy of our analysis approach and the SystemC specific coverage criteria to improve the testsuite quality.

5.1.1 SystemC Running Example

We present here an example SystemC program (Figs. 5.1 and 5.2) that will be used to showcase the main ideas of our approach throughout this section. The SystemC constructs and semantics necessary to understand the example will be explained as needed. We omitted the SystemC code required for instantiation and binding of components, i.e. the elaboration phase. The example consists of two modules *producer* and *consumer* that communicate through a FIFO. Their behavior is implemented in thread functions (producer: `prod\_thread()` Line 37—consumer: `recv()` Line 48, `filter()` Line 62, `send()` Line 72) registered in the simulation kernel. The behavior of the consumer depends on the input provided by the producer.

The FIFO provides a *write* and a *read* function that adds or removes an element, respectively. The write function is used from the producer thread in Line 41 and the read function from the consumer thread in Line 52. Both functions can potentially suspend the threads execution in case the FIFO is empty on read attempt (Line 16) or full at write attempt (Line 7). The thread becomes runnable again when the awaited event is notified in Line 11 or Line 21s, respectively

The consumer module itself consists of three threads. The thread `recv()` is responsible to retrieve the next produced element `x` (Line 53) from the FIFO and transfers it to the send thread for processing. The transfer can happen in two ways: (1) Through the `filter()` thread that applies post-processing and checking (Line 62), or (2) directly without delay to the `send()` thread for high priority items (Line 72). However, the send thread only accepts one fast transfer at a time (Line 73, controlled by the `fast\_mode` variable). Therefore, the filtering is always unconditionally initiated (Line 54) as fallback in case the send thread currently does not support fast processing. Finally, the `send()` thread will notify the `recv()` thread (Line 84) to transfer the next element.

5.1.2 Def-Use Association and Data Flow Testing

A def-use association is an ordered triple (x, d, u) such that d is a statement where variable x is defined and u is a statement where x is used. Furthermore, there is a path in the program from d to u without re-definition of x. For example, consider

```
1  struct fifo : public sc_channel {
2    fifo(sc_module_name name)
3      : sc_channel(name), num_elements(0), first(0) {}
4
5    void write(char c) {
6      if (num_elements == max)
7        wait(read_event);
8
9      data[(first + num_elements) % max] = c;
10     ++num_elements;
11     write_event.notify();
12   }
13
14   void read(char &c){
15     if (num_elements == 0)
16       wait(write_event);
17
18     c = data[first];
19     --num_elements;
20     first = (first + 1) % max;
21     read_event.notify();
22   }
23
24 private:
25   enum e { max = 10 };
26   char data[max];
27   int num_elements, first;
28   sc_event write_event, read_event;
29 };
30
31
32 SC_MODULE(producer) {
33   SC_CTOR(producer) {
34     SC_THREAD(prod_thread);
35   }
36
37   void prod_thread() {
38     wait(0); // start together with the consumer
39     const char *str = "SystemC Example"; // input to the design, can influence
           consumer behavior
40     while (*str)
41       out->write(*str++); // out is bound to fifo instance
42   }
43
44   sc_port<fifo> out;
45 };
```

Fig. 5.1 SystemC example (part 1)

```
46 SC_MODULE(consumer) {
47   SC_CTOR(consumer) { SC_THREAD(recv); SC_THREAD(send); SC_THREAD(filter); }
48   void recv() {
49     wait(0, SC_NS); // ensure send and filter are run first
50     char c;
51     while (true) {
52       in->read(c); // in is bound to fifo instance
53       x = c;
54       filter_event.notify(1, SC_NS);
55       if (x < 10) {
56         // high priortiy data handled immediately
57         send_fast_event.notify();
58       }
59       wait(recv_event);
60     }
61   }
62   void filter() {
63     while (true) {
64       wait(filter_event);
65       if (x < 0)
66         x = 0;
67       if (x > 126)
68         x = 126;
69       send_regular_event.notify(1, SC_NS);
70     }
71   }
72   void send() {
73     bool fast_mode = true;
74     while (true) {
75       if (fast_mode) {
76         wait(send_regular_event | send_fast_event);
77         fast_mode = false;
78       } else {
79         wait(send_regular_event);
80         fast_mode = true;
81       }
82       assert (x >= 0);
83       cout << x << endl;
84       recv_event.notify();
85     }
86   }
87   sc_port<fifo> in;
88 private:
89   int x;
90   sc_event recv_event, filter_event, send_regular_event, send_fast_event;
91 };
```

Fig. 5.2 SystemC example (part 2)

Fig. 5.2 variable `fast\_mode` is defined in Line 73 and used in Line 75, so it is a def-use association. A def-use association (x, d, u) is exercised by a testcase t, iff execution of t goes through definition d and then use u without re-definition of variable x in-between.

Data flow testing tries to maximize the exercised def-use associations. Essentially, it works by refining the testsuite by adding testcases until the coverage criteria are met or testing resources are exhausted. This requires to detect def-use associations and measure the data flow coverage of the current testsuite. How to do this for SystemC and thereby taking SystemC specifics into account is shown in the following. Please note that we will use the term data flow association as a generalization of a def-use association to avoid confusion. The reason is that we define a SystemC specific wait-notify association, which is also a data flow association.

5.1.3 *Data Flow Testing for SystemC*

5.1.3.1 Overview

An overview of our data flow testing approach for SystemC is shown in Fig. 5.3. Essentially, our approach combines a static and dynamic analysis to fully automatically compute a SystemC specific data flow coverage result.

The *static analysis* (upper half of Fig. 5.3) identifies the set of all-data-flow associations. Our static analysis computes an over-approximation of all def-use associations and thus also contains *infeasible* associations, i.e. associations that cannot be exercised no matter which input is applied. A precise computation of all feasible associations requires extensive use of formal verification techniques and therefore is not practical due to scalability issues. To guide test-case selection, associations are classified into different disjoint groups based on the likeliness of

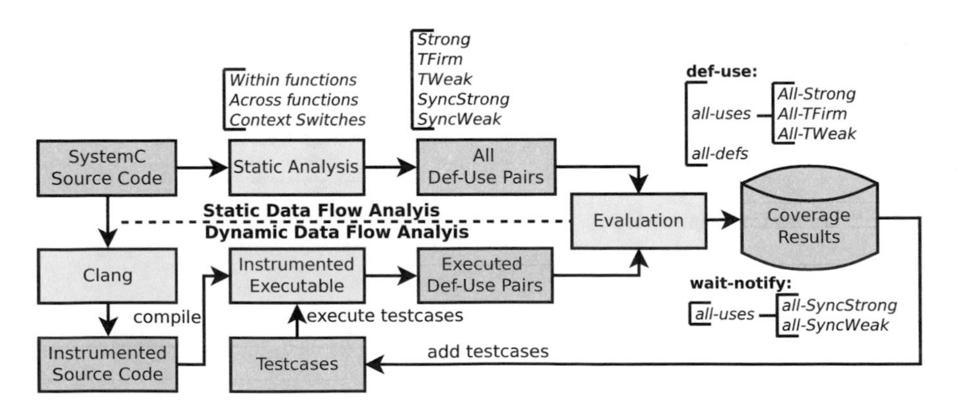

Fig. 5.3 An overview of our data flow testing approach for SystemC

being infeasible. Please note that the static analysis needs to be only run once at the beginning on the source code.

The *dynamic analysis* (lower half of Fig. 5.3) detects which data flow associations have been exercised by the current testsuite. It works by instrumenting the SystemC source code to log relevant runtime information. The instrumented source code is then compiled with a standard C++ compiler and executed for every testcase. The resulting logs are analyzed and combined to obtain the set of exercised data flow associations.

In the next step, both static and dynamic analysis results are evaluated and combined to obtain a coverage result. Essentially, the result shows which data flow associations have been exercised by at least one testcase and which have been completely missed. An association can be missed due to two reasons: (1) The testsuite is insufficient to cover it. In this case a new testcase needs to be added. (2) The association is infeasible, i.e. there is no possible input that will cover it. In this case it can be ignored.

Our classification system, that ranks associations according to their likeliness of being infeasible, allows the testing engineer to focus his efforts on promising testcases to efficiently improve the coverage result. Please note, that we do not yet consider automated test generation to exercise a specific data flow association. This is left for future work.

In the following we describe our classification system and the coverage result in more detail and demonstrate both using the running example.

5.1.3.2 Classification of Data Flow Associations

Our classification system attempts to preserve scalability of the data flow testing approach and at the same time provide meaningful results and suggestions to guide the test-case generation. We define five SystemC specific classifications: *Strong, TFirm, TWeak, SyncStrong*, and *SyncWeak*.

The first three (details see below) extend the classical notion of data flow testing that reason about variable definition and use. Therefore, these fall into the def-use association category. These classifications especially deal with the non-preemptive threads of the SystemC simulation (and hence the T in TWeak and TFirm).

The last two classifications are used to classify event-based synchronization of SystemC by means of the wait/notify function. This can also be considered as a data flow relation. The wait can be considered a definition which suspends the active thread, while the notify is considered a use. However, to avoid confusion we introduce a new data flow association called *wait-notify association* for these synchronization related flows.

Def-Use Associations Our static analysis reports the following def-use associations (x, d, u): There is a static path from d to u in the program without re-definition of x in-between. Please note, there is a static path from every context switch statement from one thread to the start of a transition of every other thread. A transition starts at the beginning of a thread and right after a context switch.

Based on this general observation we define three classifications for def-use associations (x, d, u). In this context we define a *du-path* as a static path between d and u without re-definition of x:

- *Strong* : (a) Every *du-path* is without context switch, or (b) x is a thread local variable, or (c) d is the only definition of x, i.e. x is a constant.
- *TFirm* : At least one *du-path* is without context switch and at least one *du-path* with context switch.
- *TWeak* : Every *du-path* contains a context switch.

Since a local thread variable cannot be re-defined on context switch, the def-use associations is considered Strong. This refinement of def-use associations provides clear guidelines for test-case selection. In general one should focus on Strong and TFirm associations first. In both cases there exists at least one (static) path without context switch in-between the definition and use.

Wait-Notify Associations Similarly to def-use associations, we define a wait-notify association as an ordered triple (e, w, n) where w contains a wait of event e and n contains a notify of event e, and there exists a path in the program such that w is notified from n.

This definition does not require a notify to happen after the wait during execution. This is due to timed notification, where the notify statement only schedules the notification to happen at a later point in time. Such a timed notification is much more difficult to handle precisely using static information. Furthermore, the notification can be canceled, e.g. by issuing a new notification of event e. Immediate notifications on the other hand more directly resemble the classical data flow relation as the notification happens when the notify statement is executed. Events are inherently not thread local and there is always a context switch between a wait and a notify, which makes static analysis more difficult.

Our static analysis detects the following wait-notify associations (e, w, n): (1) A timed notify n for event e is executed and the scheduled notification is not canceled until a context switch is executed. And w is a wait for event e. (2) w is a wait for event e in one thread and n is an immediate notify of event e on another thread.

This approximation can be further refined by applying an additional analysis that computes transitions between threads more precisely. However, this approximation is sufficient when dealing with SystemC synchronization primitives. One of the reasons is that often only very few wait-notify statements operate on a single event. Therefore, it is reasonable to assume that most of them are feasible.

This observation motivates to introduce the following two classifications for wait-notify associations (e, w, n):

- *SyncStrong* : Wait w should have a one-to-one relationship with a notify n for an event e.
- *SyncWeak* : Otherwise, i.e. multiple wait or notify statements are available for event e.

5.1.3.3 Coverage Result

Every classification defines a disjoint set of data flow associations. Therefore, we define a coverage criterion for each classification. For instance, the all-Strong criterion is satisfied, iff all-data-flow associations classified as Strong have been exercised. Criteria for the remaining classifications—i.e. all-TFirm, all-TWeak, all-SyncStrong, and all-SyncWeak—are defined analogously. Based on these criteria, we define specific def-use and wait-notify criteria:

- The all-uses criterion requires that all-Strong, all-TFirm, and all-TWeak are satisfied.
- The all-defs criterion requires that for every definition D in the program at least one def-use association (x, d, u) with $D = d$ is exercised.
- The all-sync criterion requires that all-SyncStrong and all-SyncWeak are satisfied.

Finally, the all-data-flow criterion is satisfied iff all-uses and all-sync criteria are satisfied.

While in general satisfying all-data-flow criterion is not practical due to imprecisions in the static analysis, which can result in infeasible data flow associations, it is possible that some of the (sub-)criteria can be fully satisfied—or at least up to a high degree, i.e. 95% of the associations have been exercised. In particular, we expect that all-Strong, all-TFirm, and all-SyncStrong to be the primarily focused criteria.

Both Strong and TFirm associations contain a (static) path without re-definition and without context switch between the definition and use. Therefore, we expect that a testcase will exercise them. Otherwise, the definition is dead code or all relevant paths (without re-definition of variable x) between definition and use are infeasible.

Similarly, if the SyncStrong criterion is not satisfied, it means that some notification has never reached a wait, or some wait has not been notified. This implies that some wait/notify statement is essentially unused.

5.1.3.4 Illustration

We have executed the SystemC example in Figs. 5.1 and (5.2) with four different inputs to gradually increase the data flow coverage. In particular, we used the following inputs for the `str` variable in Line 39: (1) "SystemC Example," (2) "Abc\x7f," (3) "a\tb\xff," and (4) "Test\x80." Table 5.1 shows the results, i.e. the statically classified data flow associations, and by which testcase they have been exercised (marked by 'X'). Infeasible data flow associations are marked with "-" for all testcases. The information is grouped in two main columns, and is read from top-to-bottom and then from left-to-right. In this order the Strong, TFirm, TWeak, SyncStrong, and SyncWeak associations are listed.

For example consider the Strong def-use association $(num\_elements, 10, 6)$ shown in the first column and exercised by all testcases. This one is Strong because

Table 5.1 Data flow associations for the SystemC example in Figs. 5.1 and 5.2, sorted by classification

Strong				
(c,5,9)	X	X	X	X
(c,57,58)	X	X	X	X
(fast_mode,80,82)	X	X	X	X
(fast_mode,84,82)	X	X	X	X
(fast_mode,87,82)	X	X	X	X
(max,25,6)	X	X	X	X
(max,25,9)	X	X	X	X
(max,25,20)	X	X	X	X
(num_elements,3,6)	X	X	X	X
(num_elements,3,15)	X	X	X	X
(num_elements,10,6)	X	X	X	X
(num_elements,19,15)	X	X	X	X
(str,39,40)	X	X	X	X
(str,39,41)	X	X	X	X
(str,41,40)	X	X	X	X
(str,41,41)	X	X	X	X
(x,72,73)			X	
(x,58,60)	X	X	X	X

TWeak				
(data,9,18)	X	X	X	X
(first,20,9)	X			
(num_elements,10,15)	X			
(num_elements,10,19)	X	X	X	X
(num_elements,19,6)	–	–	–	–
(num_elements,19,9)	X			
(num_elements,19,10)	X			
(x,72,71)	–	–	–	–
(x,72,89)			X	
(x,72,90)			X	
(x,74,71)	–	–	–	–
(x,74,73)	–	–	–	–
(x,74,89)		X		
(x,74,90)		X		
(x,58,71)	X	X	X	X
(x,58,73)	X	X	X	X
(x,58,89)	X	X	X	X
(x,58,90)	X	X	X	X

TFirm				
(data,3,18)	–	–	–	–
(first,3,9)	X	X	X	X
(first,3,18)	X	X	X	X
(first,3,20)	X	X	X	X
(first,20,18)	X	X	X	X
(first,20,20)	X	X	X	X
(num_elements,3,9)	X	X	X	X
(num_elements,3,10)	X	X	X	X
(num_elements,3,19)	–	–	–	–
(num_elements,10,9)	X	X	X	X
(num_elements,10,10)	X	X	X	X
(num_elements,19,19)	X	X	X	X

SyncStrong				
(filter_event,70,59)	X	X	X	X
(read_event,7,21)	X			
(recv_event,64,91)	X	X	X	X
(send_fast_event,83,62)				X
(write_event,16,11)	X	X	X	X

SyncWeak				
(send_regular_event,83,75)	X	X	X	X
(send_regular_event,86,75)	X	X	X	X

all paths between Lines 10 and 6 are without re-definition of num_elements and are free of context switches. In fact there is only one possible path from Line 10 to Line 6—first the write function is exited, then the while loop is not finished and so the write function is called again.

For the def-use association (*num_elements*, 10, 9), there are two paths between Line 10 and Line 9, due to the branch in Line 6. One involves a context switch and the other path does not. Therefore, this association is TFirm.

The def-use association $(first, 20, 9)$ is TWeak, because the only way to reach the use starting from the definition is through a context switch. This association is only exercised by the first testcase.

The associations involving the `max` variable are classified Strong because it is a constant, so it cannot be re-defined. The association $(c, 5, 9)$ is also Strong, even though there are two paths from 5 to 9 and one involves a context switch, because c is thread local—so it cannot be re-defined due to the context switch. Most wait-notify associations are SyncStrong because there is only a single wait and notify for the event. The ports *in* and *out* are bound in the elaboration phase, which is not shown in the example. Therefore, no def-use association is reported for them in Table 5.1.

The first testcase already achieves a reasonable data flow coverage. In particular for the FIFO component, which works independent of the actual input. Since the all-Strong and all-TFirm coverage criteria are already satisfied, the next step is to consider the TWeak associations.

The second and third inputs use special characters which are processed separately in the filter thread in Lines 65–68. With the first three inputs, all feasible def-use associations are covered. Therefore, the maximal def-use coverage has been achieved for this example. Please note that full branch and statement coverage has also been achieved with this testset.

However, the coverage with regard to wait-notify associations can still be increased. In particular the association $(send\_fast\_event, 83, 62)$ has not been exercised. The reason is that with the current testsuite all special characters were passed through the filter thread, since the send thread has not been in *fast_mode*, i.e. waiting in Line 86. Therefore, we have added a fourth testcase to exercise its association. This testcase was able to detect a bug in the design, where an invalid character is passed to the send thread.

This example demonstrates that standard def-use coverage criteria alone are not sufficient for extensive testing of SystemC designs. Our proposed SystemC specific wait-notify coverage criteria is important for a high quality testsuite.

5.1.4 Implementation Details

This section describes the static and dynamic analysis of our data flow testing framework in more details. As aforementioned, the static analysis computes an over-approximation of data flow associations classified into disjoint groups. The dynamic analysis then detects which data flow associations really have been exercised by the testsuite.

5.1.4.1 Static Analysis

The static analysis is implemented using the LibTooling library of the Clang compiler. Clang generates an *Abstract Syntax Tree* (AST) of the SystemC source code. The AST is parsed to extract the required information to do the static analysis. Then, three subsequent analysis steps are performed: (1) Local analysis within every function, (2) Information propagation across function calls, (3) Consideration of the effects of context switches.

5.1.4.2 Dynamic Analysis

The dynamic analysis works in two steps: (1) Instrument the SystemC source code using the Clang compiler framework to log data flow relevant information. Then, execute all testcases on the instrumented executable to generate the log. (2) Analyze the log line by line to build the exercised data flow associations. Both steps are described in the following.

Source Code Instrumentation In the first step the SystemC source code is instrumented to log data flow relevant information. Therefore, it is analyzed statement by statement to detect (1) definitions and uses of variables, and (2) wait and notifies of events. For every such detected information a print instruction, which writes to a log, is placed before the statement. Please note, that the order of the print instructions is important in case multiple information are available, e.g. *i++;*, in general the uses are placed before the definitions. For the while loop (and similarly the for loop) print instructions for the loop condition are replicated at the end of the loop since the condition is re-evaluated in every loop step. Therefore, it is not enough to place them only before the while loop.

5.1.4.3 Data Flow Association Construction

Def-Use Associations The def-use associations can be identified in a straightforward way. We keep a mapping of active definitions. It relates each variable to its last definition, which is updated on re-definition. Whenever a use for a variable is found in the log, the corresponding definition is retrieved from the mapping.

Wait-Notify Associations Detecting wait-notify associations requires additional work due to timed notifications. This decouples the notify statement from the actual event triggering. Therefore, we modified the SystemC kernel to write a log entry whenever an event is triggered.

For example consider the statement sequence *e.notify(1, SC_NS); wait(e);*, where an event *e* is scheduled for notification and then the thread is suspended to wait for the notification *e*. The execution log contains a notify entry for event *e* followed by a wait entry for *e*. The actual trigger from the SystemC kernel appears later. In-between can be other log entries. Other threads can also wait for event e. However, a new notification for e will cancel the one before.

Based on this information, we keep a mapping from an event to a set of active wait statements and a mapping to the last scheduled notification. Once the kernel trigger is parsed for an event e, retrieve the last notification n and set of active waits S. Then add a wait-notify association (e, w, n) for every $w \in S$. Immediate notifications are handled in the same manner, the trigger simply appears directly after the notification scheduling in the log.

5.1.5 Experimental Results

In this section we present a case study to demonstrate our DFT approach for SystemC. We consider the LEON3-based VP SoCRocket [189] which has been modeled in SystemC TLM and consider one component from it in more detail: the *Interrupt Controller for Multiple Processors* (IRQMP). IRQMP handles the interrupts coming from different connected devices using a priority mechanism. The model has I/O wires, a register file, and an APB slave interface. A total of 32 interrupts are supported. When an interrupt arrives, the corresponding bit in the register is set. The IRQMP communicates with connected processors by using an interrupt request (*irq_req*) or an acknowledgment (*irq_ack*). When an interrupt request is signaled for a processor, the IRQMP combines the `mask register` and the `pending register` with the `force register` to find the highest priority interrupt. The IRQMP also reads the `broadcast register` before forwarding the request to the processors. If the corresponding bit is set in `broadcast register`, the interrupt is broadcasted to all processors, i.e. written to the `force register` of all connected processors. In this scenario, the IRQMP expects acknowledges from all processors. On the arrival of an interrupt request, if the corresponding bit is not set in the `broadcast register`, it is simply set in the `pending register`. In this scenario, IRQMP expects an acknowledge from any processor.

The initial testsuite shipped with the IRQMP component consists of 60 tests achieving 63% statement coverage, 74% branch coverage, and 63% data flow coverage: a total number of 571 data flow associations has been computed by our static analysis, and 359 of them have been found to be exercised by the testsuite using our dynamic analysis. Initially 62% Strong, 71% TWeak, and 58% SyncWeak data flow associations are exercised. There is no TFirm association as each instruction has to pass *wait* statement, hence, a context switch is inevitable. The testsuite does not fulfill the all-uses, all-defs, and all-sync criteria. Therefore, the all-data-flow criterion is not satisfied. One interesting point is that there are no SyncStrong wait-notify pairs, instead there are 5 SyncWeak wait-notify pairs w.r.t. the only *sc_event e_signal*.

To increase the def-use coverage, the uncovered 38% Strong associations are satisfied first by adding additional tests manually. They are followed by 29% TWeak and 42% SyncWeak associations. In total, 54 additional tests are added to the initial testsuite. These new tests increased the overall data flow coverage to 88%, with 96% Strong, 91% TWeak, and 69% SyncWeak data flow associations exercised. After this point, we found it to be very hard to improve these numbers further. This demonstrates the need for future research on automated test generation techniques for SystemC that are capable to derive hard-to-find tests and to prove infeasibility of associations. With the now enhanced testsuite, we also achieve 92% statement coverage and 89% branch coverage. Despite the very high values of these code coverage metrics, they provide no insight into potential synchronization issues. In contrast our developed data flow coverage with regards to wait-notify association shows that in many cases there was no *wait* statement waiting for notify because some other function had already fulfilled the *wait* condition. This can lead to potential problems in integration of the component in a larger system/subsystem, e.g. an incoming interrupt which needs to be handled quickly can be delayed.

5.2 Verifying Instruction Set Simulators using Coverage-Guided Fuzzing

In todays design flow *Instruction Set Simulators* (ISSs) play an important role in serving as executable specification for subsequent development steps. Thus, ensuring the correctness of the ISS is crucial as undetected errors will propagate and become very costly. The ISS is an abstract SW model of a processor, typically implemented in C++ to enable a high simulation performance. Formal methods do not scale to complete verification and thus simulation-based methods are employed. They require a comprehensive testset (i.e. stimuli). Manually writing the testcases is not practical due to the significant effort required and pure random generation offers only very limited coverage.

Various approaches have been proposed to improve random generation of processor-level stimuli. Model-based test generators use an input format specification to guide the generation process and can integrate constraints processed by CSP/SMT solver [2, 29, 52]. In [121] an optimized test generation framework, by propagating constraints among multiple instructions in an effective way, has been presented. Ma et al. [147] proposed to mine processor manuals to obtain an input model automatically. Other notable approaches include coverage-guided test generation based on Bayesian networks [56] and other machine learning techniques [116].

Since the ISS is a SW model, semi-formal methods based on dynamic program analysis and constraint solving are applicable. They provide automatic ways to increase code coverage but are still susceptible to scalability issues [63, 192] or impose limitations (w.r.t. modeling memory access and loops) on the ISS [214].

Recently, in the SW domain the automated SW testing technique *fuzzing* [154] has become a standard in the SW development flow [152, 164]. Traditional fuzzing methods are *generational*, which in spirit are similar to the model-based generation techniques employed in processor-level verification and consequently have also been adopted for the verification of ISSs [149]. More recent fuzzing approaches in the SW domain employ the so-called *mutation* based technique. It mutates randomly created data and is guided by code coverage, hence avoiding the effort to create an input model. Notable representatives in this *Coverage-guided Fuzzing* (CGF) category are the LLVM-based *libFuzzer* [141] and *AFL* [9], which both have been shown very effective and revealed various new bugs [9, 141].

In this section we propose to leverage state-of-the-art CGF, as used recently in the SW domain, for ISS verification. In addition we propose a novel functional coverage metric (to complement code coverage) and mutation procedure tailored specifically for ISS verification to improve the efficiency of the fuzzer. As a case study we have implemented our CGF approach with the proposed extensions on top of libFuzzer and evaluated it on a set of three publicly available RISC-V ISSs. We found new errors in every considered ISS, including one error in the official RISC-V reference simulator *Spike*.

In the following we present our proposed CGF approach for ISS verification. Please refer to Sect. 2.3 for background information on CGF using LLVM libFuzzer as an example.

5.2.1 Coverage-Guided Fuzzing for ISS Verification

This section presents our proposed CGF approach and our fuzzing extensions tailored for ISS verification: functional coverage metric (definition in Sect. 5.2.1.2 and instrumentation to measure it in Sect. 5.2.1.3) and a specialized mutator (Sect. 5.2.1.4). Functional coverage complements code coverage and improves the thoroughness of the testset especially in detecting computational errors (which depend on operand values and structure). The use case for our mutator is to introduce specific instruction pattern into the fuzzing process, further guiding the fuzzer and allowing the fuzzer to re-use those pattern in its own mutations and cross-over operations. We start with an overview.

5.2.1.1 Overview

An overview of our CGF approach for ISS verification is shown in Fig. 5.4. Essentially, it consists of two subsequent steps: First a testset is generated by the fuzzer (upper half of Fig. 5.4), then the testset is used to verify the functionality of the ISS under test (ISS-UT, lower half of Fig. 5.4) by comparing the execution results with other reference ISSs (can be multiple). In the following we describe both steps in more detail.

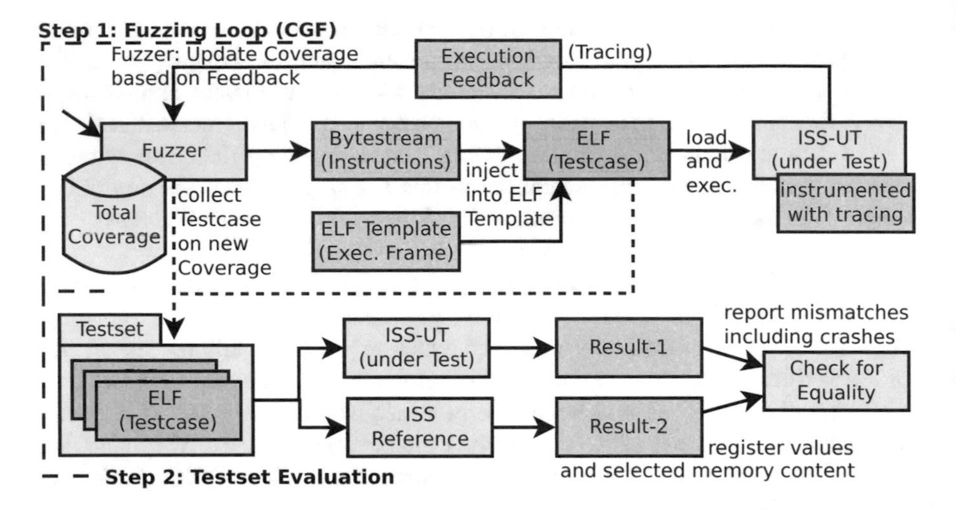

Fig. 5.4 Overview on coverage-guided fuzzing (CGF) for ISS verification

Fuzzing Loop (CGF) The fuzzer starts with an empty coverage and empty testset. The fuzzer iterates until the coverage goal or the specified time limit is reached. In each step the fuzzer generates a (binary) bytestream. We interpret this bytestream as a sequence of instructions for the ISS under test (ISS-UT). Every such bytestream is transformed into an (ELF-)testcase, by embedding the bytestream into a pre-defined ELF-template.[1]

The ELF-template contains prefix and suffix code (execution frame) that is supposed to be executed before and after the actual sequence of instructions. The prefix is responsible to initialize the ISS into a pre-defined initial state. This includes initializing all registers to pre-defined values to ensure that all ISS implementations start in the same state. The suffix is responsible to collect results and stop the simulation. It will write all register values into a pre-defined region in memory (can contain additional content beside the register values) to enable dumping the result of the execution to a file (an ISS typically provides an operation to dump specific memory regions), which can be compared.

The testcase is then executed on the ISS-UT. The ISS-UT generates execution feedback by tracing relevant information. The tracing functionality need to be instrumented into the ISS-UT. The fuzzer will analyze the execution feedback and update its coverage metrics accordingly. We consider structural and functional coverage. As already mentioned, we present our functional coverage metrics and the instrumentation to trace it in Sects. 5.2.1.2 and 5.2.1.3, respectively. In case the

[1] Technically, we use a linker script that generates an empty section in the ELF-template. Then we use *objcopy* utility with the *–update-section* argument to overwrite the empty section with the (binary) instruction bytestream.

coverage is increased by executing the testcase, the testcase is added to the fuzzer testset.

Testset Evaluation After the testset has been generated, every testcase is executed one after another on the ISS-UT and other reference ISSs. The results are compared and mismatches reported. Please note, not all mismatches are necessarily bugs. The reason is that the fuzzer is not constrained to specific well-defined subsets of the instruction set but considers all possible instructions and sequences of instructions (including illegal instructions). This behavior is intended to check specifically for uncommon (error-) cases. A common source for mismatches is differences in configuration. For example the memory size can be different or some peripheral mapped into the address space (which can not always be easily changed without intrusive modifications). A load/store instruction might then succeed for one ISS and fail for the other. We do not consider such mismatches to be bugs. Thus, mismatches need to be analyzed to check if they relate to bugs in the corresponding ISS.

5.2.1.2 Functional Coverage Metric

We define generic functional coverage metrics for registers and immediates that are applicable to a large set of ISAs. In the following we introduce functional metrics that (1) reason about the instruction structure, and (2) the operand values:

Structure Metrics A testset satisfies the coverage metric **R2** iff for every instruction with one source (RS1) and destination register (RD) at least once the case $RD = RS1$ and at least once the case $RD \neq RS1$ is observed. The coverage metric **R3** extends R2 to instructions operating on three registers (another source register RS2 is used). For R3, each of the four following cases should be observed for every such instruction:

- $RS1 = RD \wedge RS2 = RD$
- $RS1 \neq RD \wedge RS2 \neq RD \wedge RS1 \neq RS2$
- $RS1 = RD \wedge RS2 \neq RD$
- $RS1 \neq RD \wedge RS2 = RD$

R3 can be further extended for ISAs involving more source and/or destination registers.

Value Metrics The metrics **V(Rx)** and **V(Ix)** require that the Rx register and Ix immediate are assigned at least once to each special value in {MIN, -1, 0, 1, MAX}, where MIN and MAX are the smallest and largest possible value of Rx and Ix, respectively. For immediates, that are interpreted as unsigned values, e.g. shift amount, the negative values are omitted from the value set. Both metrics are applicable to every instruction having the Rx and Ix parameter, respectively. For example V(RD), V(RS1), and V(I_imm) are applicable for the RISC-V ADDI instruction (addition of register with immediate).

5.2.1.3 Instrumentation for Tracing Functional Coverage

An ISS typically consists of an execution loop that will fetch the next instruction, decode it, and then execute it. We instrument the execution step to call *begin-fcov-trace* and *end-fcov-trace* function (before/after the actual execution). Both trace functions take three arguments: (1) the instruction fetched from memory, (2) the decoded operation, and (3) a snapshot of the register values (read-only). Using two trace functions allows to capture the register state before (source register values) and after the instruction execution (destination register value). Please note, immediate values are also captured since they are encoded in the instruction itself.

5.2.1.4 Custom Mutations

A mutation is applied on a bytestream (i.e. instructions inside a testcase) and returns a modified bytestream. We propose an additional mutator tailored for ISS verification. Our mutator starts at the beginning of the bytestream and proceeds forward, applying local mutations, until the end of the bytestream is reached. In each step one of the following two mutations is performed (randomly selected):

(1) Select a random instruction and inject its opcode bits at the current position into the bytestream, overwriting existing data. This ensures that a legal instruction is used, but the parameters (source registers, destination register, immediate values, etc.) are not modified (still randomized by the fuzzer). For example the ADDI instruction of the RISC-V ISA consists of 32 bit (4 bytes) where the bits [11,7], [19,15], and [31,20] encode the parameters of ADDI: the destination register (stores the addition result), source register (first operand), and immediate (second operand), respectively. All remaining ten bits [14,12] and [6,0] encode the opcode (a fixed value) of ADDI and hence would be injected in case ADDI is selected. Optionally, a randomly selected special immediate value can also be injected into the instruction.

(2) Select a random constrained sequence of instructions and inject it into the bytestream, overwriting existing data. A sequence is a list of concrete instructions with some parameters constrained to fixed values and others randomized. A common sequence is to use two instructions to load a large constant value into a register by loading the lower and upper half separately (because typically only small immediates can be encoded in one instruction). In this case the destination register is constrained to be the same for both instructions, but the actual value is randomized (or selected from a set of special values). Sequences are defined by the verification engineer and can be specialized for each instruction set.

5.2.2 *Case Study: RISC-V ISS Verification*

As a case study we built our CGF approach on top of *libFuzzer* and verify the RV32IMA ISS extracted from our open-source RISC-V Virtual Prototype (VP) [183]. We denote this ISS as *ISS-UT*. As reference we use the following two ISS:

1. *Spike*, the official RISC-V ISA reference simulator [197].
2. *Forvis*, an ISS implemented in Haskell aiming to be a formal specification of the RISC-V ISA [58].

We have found errors in ISS-UT as well as both reference ISSs. We first describe the evaluation setting and libFuzzer integration. Then we show our detailed evaluation results.

5.2.2.1 Evaluation Setting and LibFuzzer Integration

We have instrumented ISS-UT with the *clang* compiler to trace branch coverage information. We manually (can be automated by an LLVM pass) added a call to the *begin-fcov-trace* and *end-fcov-trace* function to trace functional coverage information before/after executing one instruction, respectively. The branch coverage information is already recognized and collected by libFuzzer, and is internally mapped to unique *features* (i.e. integer values identifying the coverage information). We have extended libFuzzer to also support the proposed functional coverage information, carefully making sure to generate unique features, i.e. not interfere with the branch coverage collection.[2] Conceptually, we implement it as a chain of consecutive if-blocks matching the cases of the functional coverage metrics and mapping each case to a unique feature for every instruction, as shown in Fig. 5.5.

We generate this code automatically from a *python* script based on the RISC-V ISA specification. We also use a slightly optimized representation based on a switch-case construct to distinguish different opcodes more efficiently.

```
1 if ((op == ADDI) && (instr.RD() == instr.RS1())) features.add(1);
2 if ((op == ADDI) && (instr.RD() != instr.RS1())) features.add(2);
3 if ((op == ADDI) && (regs[instr.RS1()] == REG_MAX)) features.add(3);
4 if ((op == ADDI) && (instr.I_imm() == 0)) features.add(4);
5 // etc ...
```

Fig. 5.5 Concept of mapping functional coverage trace information to features

[2]Technically, we essentially added a new tracing collector class and extended the size of the feature representation data type. This step required adapting the fuzzer core to use sparse data structures, instead of fixed sized arrays, due to the increased feature state space.

We integrated our custom mutator into libFuzzer in a way to be randomly selected with the same probability as any of the existing mutators. As discussed in Sect. 5.2.1.4 we define instruction sequences to load a random or special value into a register and also allow to inject special immediate values.

In addition to the metrics defined in Sect. 5.2.1.2, we propose the RISC-V specific metric **R1**. R1 requires that every instruction with a destination register (RD) is executed with the case RD=0 and with the case RD≠0. This metric is useful to check that the hardwired zero register is not erroneously overwritten for some instruction. The metrics V(I_imm) and V(I_shmt) cover immediates used in computational operations.

We integrate the official RISC-V ISA tests [181] and the RISC-V Torture testcase generator [182] in the comparison to further evaluate the effectiveness of our fuzzing approach. All experiments are performed on a Linux system with an Intel Core i5 processor with 2.4 GHz and 32 GB RAM.

5.2.2.2 Evaluation Results

Table 5.2 shows the results. The table is separated into four main columns. The first column (Tests/Generator) reports which testset is used or how it has been obtained, respectively. The second column (Time) shows the time in seconds to generate (9h timeout for our fuzzer) and execute the respective testcases. Please note, the RISC-V ISA tests are directed tests that are hand-written and thus do not require a generation step. The third column shows the code—(branch coverage in particular) and functional coverage obtained by running the respective testset. The coverage is measured based on the instrumented ISS-UT. The fourth and last column shows which (and how many) errors have been found by each approach. Table 5.4 lists and describes the errors we have found in some more detail. Table 5.3 complements Table 5.2 by showing more details on the obtained functional coverage.

Already the RISC-V ISA tests can be very effective in detecting errors (Table 5.2, row T1). The testset revealed three errors (V1, V2, and V3) in ISS-UT. Furthermore, the testset is very compact and thus can be executed very fast. However, being hand-written, the testset is susceptible to miss relevant behavior, which is also reflected

Table 5.2 Evaluation results—all execution times are reported in seconds—[V1..V7] means all 7 errors V1–V7 have been found

	Time	Coverage		Errors found in ISS		
Tests/Generator	(sec.)	Code	Func.	ISS-UT	Spike	Forvis
T1: RISC-V ISA Tests	2	90.24%	37.89%	[V1..V3]	/	/
T2.1: RISC-V Torture 1000	5280	74.30%	56.20%	V1,V2	/	H2
T2.2: 5000	26,108	74.30%	57.88%	V1,V2	/	H2
T2.3: 10,000	52,168	74.30%	63.56%	V1,V2	/	H2
T3: Cov.-guided Fuzzing	32,492	100.00%	97.42%	[V1..V7]	S1	H1,H2

Table 5.3 More details on functional coverage (complements Table 5.2)

	Functional coverage (measured by instrumenting ISS-UT)							
Tests/ Generator	R1	R2	R3	V(RS1)	V(RS2)	V(RD)	V(I_imm)	V(I_shmt)
T1: RISC-V ISA tests	58.57%	61.70%	50.00%	14.29%	2.70%	7.55%	8.33%	100.00%
T2.1: RISC-V torture 1000	2.17%	66.67%	69.23%	58.82%	91.43%	52.17%	9.09%	100.00%
T2.2: 5000	2.17%	66.67%	69.23%	58.82%	91.43%	56.52%	18.18%	100.00%
T2.3: 10,000	2.17%	66.67%	69.23%	58.82%	91.43%	56.52%	63.64%	100.00%
T3: Cov.-guided fuzzing	100.00%	100.00%	100.00%	98.21%	100.00%	81.13%	100.00%	100.00%

by the obtained coverage values. Furthermore, significant manual effort is required in order to create the testset.

Torture tests reveal one additional error in Forvis (Table 5.2, rows T2.X). It can be observed that gradually increasing the testset from 1000 (Table 5.2, row T2.1) to 10,000 (Table 5.2, row T2.3) randomly generated tests does only slightly increase the coverage. The reason is that Torture receives no execution feedback, hence every test is generated independent of the previous ones.

Our fuzzer is able to detect all previously shown errors and finds six additional errors (4 in ISS-UT, 1 in Spike and 1 in Forvis), see (Table 5.2, column *"Errors found in ISS"*, row T3). Most of these errors relate to dealing with different forms of illegal instructions in different steps of the execution process. Besides being coverage-guided, a major benefit of our fuzzer is being not constrained to some specific instruction subset, as for example Torture (and hence could not detect the errors our fuzzer did, independent of the number of testcases generated). In particular the fuzzer operates on the binary level, thus it can be used to check for errors that might even be masked by a compiler/assembler (as they do not generate illegal instructions). This also enables to thoroughly check the instruction decoder unit, which even revealed an error (S1 in Table 5.4) in the RISC-V reference simulator Spike.

It can be observed that our fuzzer maximizes most coverage metrics to 100% (Table 5.2, row T3). Besides V(RS1), which is close to 100%, only the V(RD) metric is below at 81%. The reason is that the value of the destination register depends on the operand values and the operation. Some result values might even be impossible for some operations. We believe that formal methods are required to fully maximize the V(RD) metric. In total our fuzzer generates a testset with 5160

Table 5.4 Description of all errors we have found in each ISS

ISS	Error description
Spike	S1: Decoder error, which allows illegal instructions to be interpreted as *FENCE* or *FENCE.I* instruction. The reason is that not all opcode bits are present in the decoding mask.
Forvis	H1: Erroneously allows to access CSRs, representing counters for hardware performance monitoring (e.g. *cycle* CSR), from a lower privileged execution mode without first enabling the access (by setting the *mcounteren* and *scounteren* CSR).
	H2: The *REMU* instruction, which computes the remainder of a division, fails because it performs a 64 bit operation even though the ISS is configured to run in 32 bit mode.
ISS-UT	V1: Multiplication erroneously dropping the upper 32 bit of the multiplication result for large operands due to working on 32 instead of 64 bit operands.
	V2: Overwriting the hardwired *zero* register with a non-zero value (i.e. instruction with destination register RD=0 and non-zero result). Such an instruction is normally not generated by a compiler and thus can be easily missed.
	V3: CSR access instructions fail to update the CSR in case the source (RS1, contains value to be loaded into CSR) and destination (RD, will receive current value of CSR) register is the same, because in this case RS1=RD is overwritten before it is read.
	V4: Undetected misaligned branch instruction due to not four byte aligned immediates. Again, the compiler will not generate such branches and thus this error can easily be missed.
	V5: Illegal CSR instruction has side effects. A CSR access instruction reads the current CSR value into register RD and then writes the RS1 register value into the CSR. In case of a read-only CSR the write access fails and RD is not allowed to be modified.
	V6: Illegal jump instruction has side effects. The jump instruction safes the current program counter into register RD before taking the jump. However, in case the jump target address is misaligned, RD is not allowed to be modified.
	V7: Various incomplete decoder checks similar to the error were found in Spike. The reason is that ISS-UT only checks the instruction opcodes as far as necessary to uniquely identify each instruction (which is revealed by random mutations on the opcode bits).

testcases. The smallest test consists of 1 instruction and the largest of 23 instructions with an average of 3 instructions.

5.2.3 Discussion and Future Work

In our experiments we observed that our CGF approach has been very effective in finding various errors in the considered ISSs and maximized most coverage metrics to 100%. One exception is the V(RD) metric, which is below at 81%. It depends on the input parameters and thus can be very difficult to reach with simulation-based methods. To close this remaining coverage gap, we plan to investigate formal (verification) techniques at the abstraction level of VPs, e.g. [88, 100], to automatically identify inputs that will cover the missing parts or infer that the missing parts are indeed unreachable (some result values may not be possible in combination with some specific operations).

Another promising direction to complement CGF is to employ constrained random (CR) techniques. CR techniques [222] have been very effective for various system-level use cases covering both functional as well as non-functional aspects,date19: see for example [77, 91]. It is also possible to integrate CR techniques with the fuzzer, by using the CR generated testcases as input seeds for the fuzzer. This might help to guide the fuzzer towards generating longer test sequences (without illegal instructions) and at the same time reducing the number of *false-mismatches*, i.e. where two ISSs report a difference which is due to a configuration mismatch as briefly discussed in Sect. 5.2.1.1.[3] However, the fuzzer still retains the ability of generating completely random instructions (guided by the state-of-the-art fuzzing heuristics) hence the ability to cover rare corner- and error-cases (which has been very effective in our experiments) that might even be masked by a compiler/assembler. For example, the RISC-V Torture test generator only generates valid instruction sequences and thus would not be able to detect some of the errors that our fuzzer did, independent of the number of testcases generated by Torture.

Ideas from [4], to cover the values of output operands, might also be applicable in this context and help in maximizing our functional V(RD) metric. Also, recently there has been a lot of interest in integrating machine learning techniques into the fuzzing process, e.g. [65, 159], which seems very promising to investigate in our application area.

Another important direction for future work is to evaluate the effectiveness of stronger coverage metrics. Path coverage and cross coverage of functional metrics can be very effective but at the same time very challenging metrics and often impractical due to the large feature state space. Therefore, we plan to consider *selective* path coverage and (functional) cross coverage. These are only applied to selected code regions, e.g. consider all evaluation paths for each instruction separately in the ISS but not across instructions, and input operand values. This can further improve the verification result while still maintaining scalability.

Finally, we plan to broaden the scope of our evaluation to integrate further architectures and instruction sets, as well as apply our CGF approach to analyze the whole platform instead of limiting the analysis to the ISS component of the VP, and investigate RISC-V compliance testing [103, 105].

[3]We currently use automated debug scripts in case of a result mismatch that splits the input instruction sequence and repeatedly generates and executes a new ELF file on both ISSs adding one instruction after another. This allows us to pin-point the precise location of the mismatch, which in turn greatly reduces the effort to debug/analyze mismatches (in particular for longer instruction sequences).

5.3 Summary

In this chapter we proposed two approaches to improve a simulation-based verification of VPs by integrating advanced coverage-guided techniques.

First, we presented the first DFT approach for SystemC-based VPs and SystemC specific coverage criteria. The criteria use five classifications (Strong, TFirm, TWeak, SyncStrong, SyncWeak) for data flow associations. Furthermore, we explain how to automatically compute the data flow coverage results for a DUV using a combination of static and dynamic analysis. This allows to improve the coverage results by adding tests for uncovered def-use pairs since the users can get useful information. We have demonstrated the applicability in a real-world VP showing the results of one IP model.

Then, we proposed to leverage state-of-the-art *Coverage-Guided Fuzzing* (CGF) for ISS verification. We integrated a novel functional coverage metric (to complement code coverage) and mutation procedure tailored specifically for ISS verification. Our extensions complement the code coverage metric used by CGF and thus improve the efficiency of the fuzzing process. We have implemented our CGF approach with the proposed extensions on top of the LLVM-based libFuzzer and in a case study evaluated it on a set of three publicly available RISC-V ISSs. We observed that our fuzzer has been very effective in maximizing most coverage metrics and in finding various errors. Fuzzing is particularly useful to trigger and check for corner- as well as error-cases and can complement other test-case generation techniques. We found new errors in every considered ISS, including one error in the official RISC-V reference simulator *Spike*. In our discussion we sketched various directions for future work to further improve and complement our CGF.

Chapter 6
Verification of Embedded Software Binaries using Virtual Prototypes

After verifying the VP, the VP is used as platform for SW development. Todays SW is becoming increasingly complex and encompasses several abstraction layers ranging from bootcode and device drivers to complex libraries and operating systems. Verification of the embedded SW is very important to avoid errors and security vulnerabilities.

This chapter presents two novel VP-based SW verification approaches to improve the VP-based design flow. In a standard VP-based design flow SW is primarily verified using simulation-based methods that employ directed or constrained random generated testcases. However, this standard verification flow either requires significant manual effort for test-case generation and refinement or suffers from limited coverage results. In addition, it is highly susceptible to miss corner case errors.

In this chapter two approaches are presented that combine concolic testing and *Coverage-guided Fuzzing* (CGF) with VPs to enable a sophisticated and comprehensive test-case generation for embedded SW. This integration is challenging due to complex HW/SW interactions and peripherals. These proposed approaches improve the existing VP-based SW verification flow by integrating stronger coverage metrics and providing automated test-case generation techniques by leveraging advanced mutation-based heuristics and formal methods.

Section 6.1 presents the first proposed approach for concolic testing of binaries targeting RISC-V systems with peripherals. The approach works by integrating the *Concolic Testing Engine* (CTE) with the architecture specific *Instruction Set Simulator* (ISS) inside of a VP. A designated CTE-interface is provided to integrate (SystemC-based) peripherals into the concolic testing by means of SW models. This combination enables a high simulation performance at binary level with comparatively little effort to integrate peripherals with concolic execution capabilities. The approach has been effective in finding several buffer overflow related security vulnerabilities in the FreeRTOS TCP/IP stack. The approach has been published in [101].

V. Herdt et al., *Enhanced Virtual Prototyping*,
https://doi.org/10.1007/978-3-030-54828-5_6

Section 6.2 presents the second proposed approach that leverages state-of-the-art CGF methods in combination with SystemC-based VPs for verification of embedded SW binaries. Using VPs, the approach allows a fast and accurate binary-level SW analysis and enables checking of complex HW/SW interactions. To guide the fuzzing process the coverage from the embedded SW is combined with the coverage of the SystemC-based peripherals. The experiments, using RISC-V embedded SW binaries as examples, demonstrate the effectiveness of the proposed approach. The approach has been published in [107].

6.1 Concolic Testing of Embedded Binaries

In the SW domain the automated concolic testing technique has been shown very effective. Essentially, concolic testing successively explores new paths through the SW program by solving symbolic constraints that are tracked alongside the concrete execution. This combination of concrete with symbolic execution enables an efficient exploration of a large set of different program paths.

A full embedded system solution consists of SW and HW. The traction coming from open-source SW recently reached open-source HW. In particular, *RISC-V* has started to become a game changer for IoT processors. RISC-V is an open and free *Instruction Set Architecture* (ISA) [218]. Since 2015 the RISC-V ISA standard is maintained by the non-profit RISC-V foundation [180] which has more than 200 members aiming innovation. However, to the best of the authors knowledge, concolic testing of binaries targeting RISC-V systems with peripherals has not been considered yet.

This section presents such a concolic testing approach for binaries targeting RISC-V systems with peripherals.[1] The proposed approach works by integrating the *Concolic Testing Engine* (CTE) with the architecture specific *Instruction Set Simulator* (ISS) inside of a *Virtual Prototype* (VP). A designated *CTE-interface* is provided to integrate additional peripherals through a SW-library that is linked with the RISC-V binary under test into a combined RISC-V binary. The *CTE-interface* consists of a small set of interface functions, tailored for SystemC-based peripherals with TLM 2.0 communication [71, 113]. This approach enables a high simulation performance, by tightly integrating the concolic execution engine with the VPs ISS for the specific target ISA, and at the same time significantly reduces the implementation effort in adding concolic execution capabilities to each peripheral (since peripherals are executed as SW on top of the VP and hence *inherit* the concolic execution capabilities of the VP). The experiments demonstrate the efficiency and applicability of the approach in analyzing real-world embedded

[1]Please note, the RISC-V ISA is used as a case study. The proposed methodology is applicable to other ISAs as well.

applications. Several buffer overflow related security vulnerabilities have been found in the FreeRTOS TCP/IP stack.

Most of existing work on concolic testing and symbolic execution has been focusing on testing non-embedded SW that does not or only rarely interacts with HW. KLEE [27] and SAGE [64] are among the pioneering work that has made the techniques really work for real-world SW. While KLEE operates on the LLVM intermediate representation and requires the source code, SAGE works directly on the x86 binary. Operating on the binary level is important to achieve accurate verification results, as binaries are code that will actually be deployed. Thus, despite being more difficult due to the complexity of lower-level constructs, concolic testing of binaries (mostly x86 and ARM to some extent) is increasingly being considered by subsequent work, e.g. S2E [34], Mayhem [30] or Angr [194]. More recently, concolic/symbolic testing has gained attention from the HW verification community, see e.g. [5, 171] for RTL and [88, 100, 142] for SystemC VPs. The symbolic simulation presented in Chap. 4 also falls into this category.

For embedded SW, these approaches are not sufficient due to the much stronger dependence of the SW on the underlying HW. This gave rise to a number of specialized HW/SW symbolic (concolic) co-validation approaches that are related to the approach presented in this chapter. They mainly differ on how the underlying HW is being integrated. Horn et al. [110] and Ahn and Malik [6] use virtual peripheral models manually extracted from QEMU. Mukherjee [157] on the other hand integrates HW Verilog models. Recently, [111] proposed a new formalism called instruction level abstraction to formally model SW-visible behavior of HW. This enables more scalable symbolic exploration but it is unclear whether the abstraction can be fully automated. These approaches operate at the source code level. FIE [41] is more similar to our approach in the sense that it targets SW running on a specific HW platform. FIE brings a set of modifications to KLEE to resemble an MSP430-based execution environment and optimizations for more scalable concolic testing (e.g. memory smudging and state pruning). However, it still requires the SW source code and cannot handle inline assembly instructions that often can be found in low-level embedded SW. Therefore, Inception [39] introduces an assembly to LLVM-IR lifting approach. AVATAR [223] extends S2E to allow hybrid binary concolic testing with physical devices.

Note that while we describe our approach in some details, we do not claim the concolic testing technique to be the central contribution. Rather it is the combination on how the concolic execution engine is integrated with the VP with the method for integration of (SystemC-based) peripherals. By this means a practical framework is provided filling the gap of concolic testing for embedded RISC-V binaries that is not possible with any existing framework. Conceivably, Angr—operating on the VEX intermediate representation at binary level—can be extended to support RISC-V instructions. However, due to its focus on non-embedded SW, extending Angr to the same extent of capability of handling RISC-V binaries and peripherals would require a lot of effort. Furthermore, based on its promises, RISC-V deserves a tailored platform that is more lightweight and amenable to specific optimizations.

6.1.1 Background on Concolic Testing of SW

Overview Concolic testing is a technique to successively explore new paths through the SW program. It uses concolic values in place of pure concrete ones. A concolic value is a pair with a concrete N and a symbolic part x written as (N, x). The concrete part is always available. We denote a concolic value to be *symbolic*, if the symbolic part x is also available. Otherwise ($x = /$) we call it *concrete*. For concolic testing, input variables (or memory regions in general) are marked to be symbolic. Furthermore, a (symbolic) *Path Condition* (PC) and a list of (symbolic) *Trace Conditions* (TCs) are tracked. After each execution, an SMT solver checks each emitted TC for satisfiability. Each satisfiable TC represents a new testcase (input). An input assigns concrete values to symbolic variables (i.e. memory regions), thus preventing them being assigned random values and guiding exploration towards a specific path.

Concolic testing starts by (concretely) executing a random path through the SW program (i.e. all variables marked symbolic are simply assigned random values). During SW execution the symbolic input constraints are propagated alongside the concrete execution. Let us consider an example operation combining two operands A and B. In case one operand $A = (1, x_0)$ has a symbolic value and the other $B = (2, /)$ has not B will be first converted to $(2, 2_S)$, where 2_S is an SMT expression representing the constant value 2. Thus, for example $A + B$ would then result in $(3, x_0 + 2_S)$. In the following we often omit the S suffix in case the context is clear and write 2_S simply as 2.

At each branch instruction with symbolic condition $s = (c, x)$, the path condition PC is extended and a trace condition TC_i is emitted, either: (1) $TC_i = PC \land \neg x$ and then $PC := PC \land x$ in case the branch is taken (c evaluates to *true* in the concrete execution), or (2) $TC_i = PC \land x$ and then $PC := PC \land \neg x$, otherwise. For verification purposes, *assume(s)* adds x to the current PC to prune irrelevant paths and emits a TC in case c is *false*. *Assert(s)* checks c and emits a $TC = PC \land \neg x$ in case c is *true*. Then PC is extended with x.

Concretization Symbolic values (N, x) can be concretized, i.e. converted to concrete values $(N, /)$, by adding the constraint $N = x$ to the path condition PC. This can be useful in case parts of the simulation only support concrete values. Optionally, trace conditions can be emitted to (try) generate different concrete values N (e.g. the minimum and maximum possible values would be good candidates).

6.1.2 Concolic Testing Engine for RISC-V Embedded Binaries

CTE enables concolic testing of binaries targeting RISC-V systems with peripherals. We start with an overview of CTE. Then, we present our approach on integrating peripherals into CTE in more details and finally show an example that illustrates how CTE interoperates with SW in combination with peripherals.

6.1.2.1 Overview

Figure 6.1 shows an overview on the architecture of CTE. Essentially, CTE consists of three parts:

1. A SW-library that contains *CTE-interface* functions and a set of peripheral models. The CTE functions allow to declare and reason about symbolic variables and enable CTE to interface with peripheral models.
2. A VP that is operating with concolic, instead of concrete, data types. The VP essentially consists of an ISS that executes instructions one after another and a memory that stores code and data. Both, ISS and memory use concolic data types and propagate symbolic constraints during program execution. The ISS supports the RISC-V RV32IMC ISA. Additional peripherals are integrated into the simulation through the CTE SW-library.
3. A concolic exploration engine that is successively exploring new paths through the SW program (binary) by leveraging the VP and solving constraints on symbolic variables using an SMT solver.

Initialization and Exploration CTE of SW works as follows: First, the SW is compiled together with the CTE SW-library into a RISC-V ELF (binary file). Variables, representing input data, are marked to be symbolic (using the CTE-interface functions). Then, the (concolic execution) VP is instantiated and the (combined) RISC-V ELF is loaded into the VPs memory. The exploration engine then starts by assigning each symbolic variable a random value (empty input) and then successively generates new inputs (based on the observed constraints) to explore different paths through the SW program. In particular it will try to generate new inputs in case of a symbolic branch condition or *assume* or *assert* function. The VP is cloned each time before executing a new input to preserve the initial VP state and allow exploration of different paths. CTE continues until all inputs have been processed or a runtime check fails. CTE checks for SW assertion violations as well as some generic error sources: null pointer dereference, access (read or write)

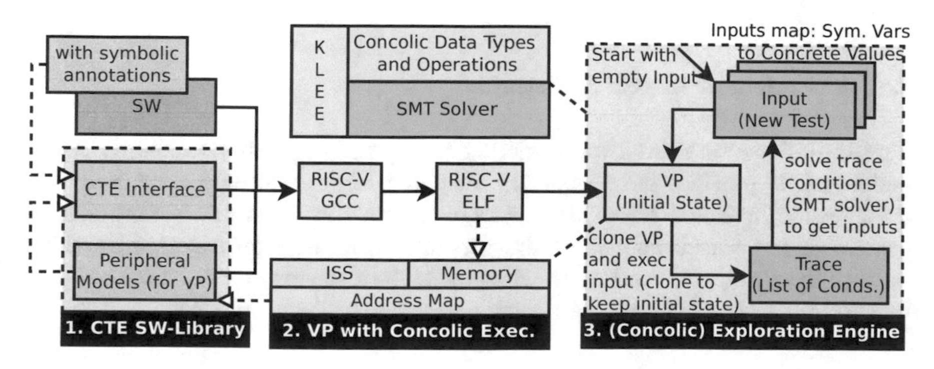

Fig. 6.1 Concolic Testing Engine (CTE) architecture overview

of an illegal memory address and invalid jump target (either not properly aligned or a jump to an illegal instruction). In our experimental evaluation we also check for overflows of heap allocated memory in FreeRTOS (more details will follow later).

Peripheral Integration To implement CTE for embedded binaries (targeting a specific ISA) with peripherals, there are in general two fundamental choices with different trade-offs between execution performance and implementation effort: (1) Integrate concolic execution into every component of the VP, i.e. including every peripheral. This fully specialized solution requires significant effort (for each peripheral) but leads to the best performance. (2) Use a generic (x86) symbolic/-concolic execution engine like S2E and run the VP, which in turn simulates the embedded binary, inside of S2E. This generic solution requires only little integration effort but has a significant performance overhead, in particular due to running the (VPs) ISS inside another simulator (S2E).

We aim to combine the benefits of both approaches and alleviate their respective major disadvantages. Our approach enables a high simulation performance, by tightly integrating the concolic execution engine with the VPs ISS for the specific target ISA (RISC-V in our case). This avoids the (very significant) additional interpretation layer in the ISS, which arguably is the component with the greatest performance impact. At the same time we integrate peripherals into the VP by providing SW models of the peripherals. Thus, peripherals are compiled to the RISC-V instruction set and normally executed on our VP alongside the actual binary that is tested. This has the major benefit, that the peripherals are executed with concolic data types and hence propagate symbolic constraints without further modification. While this involves a small (currently manual) transformation step to obtain an appropriate peripheral SW model (we use an existing SystemC model as starting point), it is much less effort than integrating concolic execution capabilities directly into the (SystemC-based) peripheral. Hence, our approach enables a fast simulation performance with comparatively little implementation effort.

In the following we start with an example that demonstrates our main modeling concepts for peripherals and then present how peripherals are accessed from SW and interoperate with the concolic exploration engine.

6.1.2.2 Peripheral Modeling Concepts

For illustration of our peripheral modeling concepts, Fig. 6.2 shows a simple sensor peripheral. It provides three (memory mapped) 32 bit registers: *data* holds the most recent generated sensor data, *scaler* controls how fast new sensor data is generated, and *filter* is applied on the generated data. Register access is provided through the *transport* function, which roughly corresponds to a TLM *b_transport* function.[2] New sensor data is periodically generated by the *update* function (Line 8).

[2]It is easily possible to write a *transport* wrapper that stores the arguments in a *tlm_generic_payload* and then calls the real *transport* function to avoid modifying the peripheral.

```
1  static uint32_t scaler = 25;
2  static uint32_t filter = 0;
3  static uint32_t data = 0;
4  #define SCALER_REG_ADDR 0x00
5  #define FILTER_REG_ADDR 0x04
6  #define DATA_REG_ADDR 0x08
7
8  void update() {
9    // overwrite data with new concolic
         bytes
10   CTE_make_symbolic(&data,
         sizeof(data), "d");
11   CTE_assume(data >= MIN_SENSOR_VALUE
         && data <= MAX_SENSOR_VALUE);
12   data -= filter;
13
14   // PLIC receives interrupts,
         prioritize them and calls a
         CTE-function to notify the VP
15   plic_process_interrupt(2
         /*IRQ_NUMBER*/);
16
17   // corresponds to simple thread
         wait in SystemC (or just a
         method process)
18   CTE_notify(&update,
         scaler*CYCLES_PER_MS);
19   CTE_return();
20 }
21
22 // corresponds to a simple TLM
       transaction
23 void transport(uint32_t addr, uint8_t
       *data, uint32_t size, _Bool
       is_read) {
24   CTE_assert(size == 4);//only
         access whole reg.
25   uint32_t *vptr = (uint32_t *)data;
26   uint32_t *reg = 0;
27
28   // pre-process actions
29   if (addr == SCALER_REG_ADDR) {
30     if (!is_read)
31       CTE_notify(&update,
             scaler*CYCLES_PER_MS);
32     reg = &scaler;
33   } else if (addr == DATA_REG_ADDR) {
34     reg = &data;
35   } else if (addr == FILTER_REG_ADDR)
         {
36     reg = &filter;
37   } else {
38     assert(0 && "invalid addr");
39   }
40
41   if (is_read) *vptr = *reg;
42   else *reg = *vptr;
43
44   // post-process actions
45   if (addr == FILTER_REG_ADDR &&
         !is_read)
46     if (filter >= MIN_SENSOR_VALUE)
47       filter = MIN_SENSOR_VALUE+1;
48
49   CTE_return();
50 }
```

Fig. 6.2 Simple sensor peripheral that illustrates the concept on peripheral modeling in combination with the CTE-interface

Therefore, the data register is overwritten with a new symbolic value (Line 10) and constrained to stay in the sensor range (Line 11). An interrupt is triggered by calling the *PLIC* peripheral (a RISC-V specific *Platform Level Interrupt Controller*, implementation not shown) in Line 15. PLIC will prioritize incoming interrupts and eventually notify the ISS using the *CTE_trigger_irq* function. *CTE_notify* instructs the VP to call the function (address provided as first argument) after the specified number of execution cycles has passed (second argument). In case the function already has a pending notification, it will be reset. To handle notifications, the VP has a simple timing model that assigns each RISC-V instruction a fixed number of cycles, which are added up during simulation. *CTE_notify* allows to model simple

An optional *delay* parameter could also be passed in and returned back to the VP through the CTE-interface.

```
1  uint32_t*SENSOR_SCALER_REG_ADDR =
       (uint32_t*)0x10000000;
2  uint32_t*SENSOR_FILTER_REG_ADDR =
       (uint32_t*)0x10000004;
3  uint32_t*SENSOR_DATA_REG_ADDR =
       (uint32_t*)0x10000008;
4
5  volatile _Bool sensor_has_data = 0;
6  void sensor_irq_handler() {
7    has_sensor_data = 1;
8  }
9
10 int main() {
11   register_interrupt_handler(
         2/*IRQ_NUMBER*/,
         sensor_irq_handler);
12   uint32_t filter;
13   CTE_make_symbolic(&filter,
         sizeof(filter), "f");
14   *SENSOR_FILTER_REG_ADDR = filter;
15   *SENSOR_SCALER_REG_ADDR = 50;
16
17   while (!sensor_has_data) { // check
         for sensor
18     asm volatile ("wfi"); // wait for
           any irq
19   }
20   uint32_t n = *SENSOR_DATA_REG_ADDR;
21   CTE_assert (n <= MAX_SENSOR_VALUE);
22   return 0;
23 }
```

Fig. 6.3 Example SW that is accessing the sensor peripheral (shown in Fig. 6.2)

processes. Currently, we only support concrete delay arguments to *CTE_notify*. In case a symbolic argument is used, it will be concretized. *CTE_return* returns execution control back to the SW program, which is interrupted in order to enter peripheral functions (*transport* and *update*). We present more details on this in the following (sub-)section. Finally, we provide the *CTE_get_cycles* function to obtain the current number of simulation cycles from the ISS to model the RISC-V specific *CLINT* (Core Local INTerruptor) peripheral. *CLINT* triggers periodic (configurable) timer interrupts and allows the SW to obtain the current simulation time.

Software Side Figure 6.3 shows an example SW that is accessing the sensor peripheral (shown in Fig. 6.2). The SW first installs an interrupt handler to listen for the sensor interrupt (Line 11). Then, it initializes the sensor (Lines 14–15) using memory mapped I/O and waits for the sensor interrupt (which indicates new data). The *filter* register is assigned a symbolic value, while *scaler* is assigned a concrete one. Finally, the SW retrieves the sensor data value and checks its validity (Lines 20–21).

To handle memory mapped accesses, the VP has an address map that contains specific non-overlapping address ranges for the memory and every peripheral. This address map information is obtained from a configuration file. Every memory access transaction, i.e. RISC-V *load/store* instruction, is matched against the address ranges and then routed accordingly. This step involves a *global-to-local* address translation. For example executing a RISC-V SW (store word) instruction with address 0x10000004 (Line 14 in Fig. 6.3), and assuming the sensor is mapped to e.g. the range (0x10000000, 0x1000ffff), then the access address is translated to 0×4 and routed to the sensors *transport* function. This step involves a context switch to the peripheral SW and then back to the main SW. In the following we provide more details on this step.

$$I_0\colon \square \xrightarrow{\text{F}=(31,f_0)} \text{if(F} \geq 16) : T \xrightarrow{\text{F}=(17,/),\text{D}=(4,d_0)} \text{assume}(D \geq 16 \wedge D \leq 64) : F$$

$$I_2 : \square \xrightarrow{\text{F}=(23,f_0)} \text{if(F} \geq 16) : T \xrightarrow{\text{F}=(17,/),\text{D}=(45,d_0)} \text{assume}(D \geq 16 \wedge D \leq 64) : T$$

$$\xrightarrow{\text{D}=(28,d_0-17)} \text{assert}(D \leq 64) : T$$

$$I_3 : \square \xrightarrow{\text{F}=(21,f_0)} \text{if(F} \geq 16) : T \xrightarrow{\text{F}=(17,/),\text{D}=(16,d_0)} \text{assume}(D \geq 16 \wedge D \leq 64) : T$$

$$\xrightarrow{\text{D}=(\text{UINT\_MAX},d_0-17)} \text{assert}(D \leq 64) : F$$

Fig. 6.4 Example concolic execution paths through the SW and peripheral shown in Figs. 6.3 and 6.2

Context Switching: SW and Peripheral To switch execution context from the SW to a peripheral function *FN*, the VP (actually the ISS) saves its current execution context (i.e. register values and program counter) to an internal stack. Then, the program counter is simply set to the address of *FN*. To return, the previously stored execution context is restored. Therefore, we introduced the *CTE_return* function.[3] Using a stack to save the execution context allows peripherals to access other peripherals memory through a memory mapped access.

Arguments between the VP and the SW peripherals are passed through registers and a (transaction) data array placed in memory (to hold larger arguments). The *transport* function takes four arguments: the (local) access address, the number of bytes to access, a data pointer (*uint8_t**), and a Boolean flag indicating the access type (read/write). The data pointer is set up to point to the data array. These four arguments are stored in the VP registers (*a0–a3*, following the RISC-V function calling convention) prior to switching context to the transport function. The address of the *transport* function for each peripheral as well as the data array is obtained (during VP initialization) by parsing the ELF symbols (i.e. given a name of a symbol, it is possible to obtain its address).

6.1.2.3 Concolic Testing Example

This section illustrates how CTE interoperates with SW and peripherals. Figure 6.4 shows the relevant execution paths for our sensor SW example. It shows the input, updates to the *filter* (F in Fig. 6.4) and *data* (D in Fig. 6.4) variables, and instructions generating trace conditions.

CTE starts without input constraints ($I_0 = \{\}$), thus assigning each symbolic variable a random value and therefore exploring a random path through the SW

[3]Though, in general it is also possible to automatically infer the end of *FN* by e.g. monitoring function calls and returns.

program. For this first run the assumption is that *filter* and *data* are (randomly) assigned the concolic values $(31, f_0)$ and $(4, d_0)$ in Lines 13 and 10, respectively. Furthermore, for this example the constants *MIN_SENSOR_VALUE=16* and *MAX_SENSOR_VALUE=64* are defined. The first trace condition $TC_1 = \neg(f_0 \geq 16)$ is generated by the branch execution in Line 46, due to the memory mapped write of the symbolic filter value in the SW (Line 14). The second trace condition $TC_2 = (f_0 \geq 16) \wedge (d_0 \geq 16 \wedge d_0 \leq 64)$ is generated by the *assume* function in Line 11, due to the symbolic data value, while the main SW waits for an interrupt (Line 18). Thus, the *assume* function evaluates to *false* and this path ends.

Both trace conditions are satisfiable, resulting in e.g. the two new inputs $I_1 = \{f_0 = 1, d_0 = 25\}$ and $I_2 = \{f_0 = 23, d_0 = 45\}$ (obtained by using the SMT solver), respectively. It depends on the search strategy of the exploration engine, which input is considered next. For this example, we continue with I_2. Until the *assume* function (Line 11) it follows the same path as the initial execution. However, different concrete values are assigned at the *CTE_make_symbolic* functions based on I_2. Thus, this time the *assume* function evaluates to *true* and execution continues. Please note, no trace conditions have been generated until now to avoid re-exploration of the first execution path (and its descendants, i.e. I_1). Next, *filter* is applied to the generated *data* (Line 12) and an interrupt is triggered (Line 15). The main SW will leave the interrupt waiting loop, fetch the data register value (Line 20), and check the assertion in Line 21. The concrete part of the condition evaluates to *true*, the assert generates a trace condition $TC_3 = (f_0 \geq 16) \wedge \neg(d_0 - 17 \leq 64)$, and this path ends. TC_3 is satisfiable with e.g. $I_3 = \{f_0 = 23, d_0 = 16\}$. With I_3 the data register in Line 12 will underflow and thus the assert in Line 21 is violated. This is due to an incorrect filter value assigned in Line 47. It should use *minus one* instead of *plus one*. This example also shows that it is important to integrate the peripheral logic into the testing approach to avoid missing relevant behavior of the overall system.

6.1.3 Experiments

We have implemented our proposed CTE for concolic testing of binaries targeting RISC-V systems with peripherals. As symbolic back-end we use KLEE v1.4.0 with STP solver v2.3.1. We evaluate CTE in two steps: First, we evaluate the performance of our approach in Sect. 6.1.3.1. Then, we apply CTE to test the FreeRTOS TCP/IP stack. Our approach has been effective in finding various security vulnerabilities.

6.1.3.1 Performance Evaluation

We evaluate the performance of our CTE by comparing against: (1) a generic symbolic execution solution using S2E, and (2) a VP supporting RISC-V binaries [96]. The ISS in VP is similar to that in CTE but only supports concrete execution.

Hence, VP uses DMI for memory access and native (C++ uint32_t) data types for instruction execution (instead of a custom concolic data type that propagates a concrete and symbolic value for each instruction). Furthermore, VP is using normal SystemC peripherals. For the S2E solution, we run VP simulating the RISC-V binary on top of S2E, thus allowing symbolic execution of RISC-V binaries with little implementation overhead (only a small interface layer is required to propagate the *make_symbolic*, *assume*, and *assert* functions from the RISC-V binary through the VP into S2E).

Table 6.1 shows the results. All runtimes are provided in seconds. It is separated into 11 columns. The columns show, in order from left to right, the benchmark name (column: Benchmark), number of instructions executed (column: #instr), number of lines (LOC) in C and RISC-V assembly (columns: C and ASM), the simulation time in seconds on the VP, S2E, and our CTE (columns: VP, S2E, and CTE), and relevant statistics for our CTE: factor of improvement (FoI) compared to S2E (column: FoI S2E), time in seconds spent with solver queries (column: stime), number of different paths explored (column: #paths), and number of solver queries (column: queries).

Table 6.1 is separated into two halves. The upper half shows benchmarks that only operate on concrete values, while the lower half shows benchmarks including symbolic values (hence exploring multiple execution paths). For symbolic benchmarks, the reported number of instructions is combined over all paths. As benchmarks we use *qsort* from the *newlib* C library, a standard *dhrystone* implementation, *sha512* performs a checksum computation, *counter* involves counting related constraints, *fibonacci* uses a recursive implementation (function call intensive), and *freertos-sensor* embeds a sensor application into FreeRTOS tasks. The */s* name suffix denotes that the benchmark uses symbolic values.

Comparing our CTE approach with VP, it can be observed that the overhead of using concolic execution is around a factor of 2.2x. The overhead of running the VP inside of S2E is between 21x and 49x compared to CTE on our benchmark set. The main reason is the additional interpretation overhead in the ISS component of the VP. With symbolic values, the overhead of S2E compared to CTE is even more pronounced. Speed-ups of multiple orders of magnitude can be observed. The (symbolic) state space representation is more heavy-weight and complex in S2E. Furthermore, the additional abstraction layer leads to increased number of forks in the symbolic execution engine. In one case (*freertos-sensor/s*) we observed only an improvement of 10.9x. The reason is that CTE explores all paths from the beginning instead of forking an active execution. This leads to significant overhead in re-initializing the FreeRTOS memory image. This inefficiency can be solved by cloning the CTE execution state after initialization is finished and then start the exploration engine from this point. Our evaluation shows that an ISA specific symbolic execution engine can provide significant speed-ups compared to a generic symbolic execution engine (in particular optimizations of the ISS are important). At the same time, our approach requires considerable less implementation effort compared to a fully specialized symbolic execution engine, by integrating SW models of peripherals instead of adding symbolic execution capabilities to every peripheral.

Table 6.1 Experiment results—timeout (T.O.) set to 7200 s (2 h)

		LOC		Sim. Time (s)			CTE Statistics			
Benchmark	#instr	C	ASM	VP	S2E	CTE	FoI S2E	stime	#paths	#queries
qsort	52,343,639	212	1126	1.83	122.33	3.87	31.6x	/	1	/
sha512	75,997,581	175	1132	2.71	136.90	4.62	29.6x	/	1	/
dhrystone	238,000,584	273	515	8.42	421.60	19.70	21.4x	/	1	/
freertos-sensor	21,348,342	217	18,230	1.35	107.37	2.19	49.0x	/	1	/
counter/s	642,710	31	79	/	1264.21	41.74	30.3x	37.82	452	904
fibonacci/s	683,970	37	170	/	394.14	1.90	197.5x	0.15	22	41
qsort/s	58,648	219	1192	/	T.O	32.90	>218x	32.58	121	600
freertos-sensor/s	301,697,668	241	18,329	/	422.78	38.66	10.9x	1.60	463	968

6.1.3.2 Testing the FreeRTOS TCP/IP Stack

To further evaluate the effectiveness of our tool on large embedded SW we have analyzed the TCP/IP stack of FreeRTOS v10.0.0 in combination with the RISC-V port of the FreeRTOS kernel. Essentially, we inject a single (small) packet with symbolic size and content into the TCP stack and check for generic execution errors (including FreeRTOS assertions) and heap buffer overflows.

Test Setup IP packets are processed in FreeRTOS inside the platform independent *IP-task*. The IP-task requires a driver to receive (send) packets from (to) the network card. The FreeRTOS documentation provides a generic porting guide [173] to create a new driver and connect it with the IP-task. Using the generic driver code, it is only necessary to implement three additional glue functions: (1) ReceiveSize, (2) ReceiveData, and (3) SendData.

Essentially, to receive a packet, the driver waits for a network card interrupt. Then, it asks the network card for the number of received bytes (*ReceiveSize* function), allocates a buffer on the heap large enough to hold *ReceiveSize* bytes, and asks the network card to store the data in the buffer (ReceiveData function). Finally, the allocated buffer is delivered to the (platform independent) IP-task for further processing. We model the network card as a peripheral (as described Sect. 6.1.2.2). Inside the peripheral we store a packet buffer of 512 bytes with symbolic content. Furthermore, we define a symbolic integer variable N and add the assumption $N \leq 512$ in the peripheral. The driver glue functions access our peripheral using memory mapped I/O. Our model simply ignores outgoing packets, returns N when asked for *ReceiveSize*, and copies N bytes into the data pointer provided by *ReceiveData*.

In the (SW) main function, we create and initialize the FreeRTOS network (IP-task) and driver processing task and start the FreeRTOS scheduler (available from the FreeRTOS kernel). Then we create and bind a TCP socket in *listening* mode to ensure that the IP-task does not drop TCP packets (which is the case if no initialized TCP socket is available). Finally, we start the actual symbolic testing by calling the *init* function of our peripheral, which in turn will trigger an interrupt notifying the driver to receive and process a packet. Interrupts are processed using our *PLIC* peripheral. Timer interrupts for the FreeRTOS scheduler processing are generated using our *CLINT* peripheral.

Please note, the IP-task is non-terminating and will wait for new packets from the driver indefinitely. Thus, we added a switch to stop simulation after one packet has been processed, to allow CTE to explore other paths as well (alternatively we could bound the search depth in CTE).

Heap Buffer Overflow Detection To check for heap buffer overflows, we provide wrappers for the FreeRTOS memory management functions *pvPortMalloc* (allocate memory) and *vPortFree* (free memory). We use the linker options *-Wl,–wrap=pvPortMalloc -Wl,–wrap=vPortFree* with GCC to automatically redirect all accesses of *pvPortMalloc* to our wrapper function *__wrap_pvPortMalloc*. The original function is available as *__real_pvPortMalloc* (*vPortFree* handled in same way).

```
1  #define PROT_ZONE_SIZE 512 //in bytes
2
3  void *__wrap_pvPortMalloc( size_t
       xWantedSize ) {
4    size_t xSize = xWantedSize +
         2*PROT_ZONE_SIZE;
5    // call the real FreeRTOS
         pvPortMalloc function
6    uint8_t *p = (uint8_t*)
         __real_pvPortMalloc(xSize);
7    if ( p == NULL )
8      return NULL;
9    void *addr = (void *)(p +
         PROT_ZONE_SIZE);
10   CTE_register_protected_memory(
         addr, xWantedSize,
         PROT_ZONE_SIZE );
11   return addr;
12 }
13
14 void __wrap_vPortFree( void *pv ) {
15   CTE_assert( pv != NULL );
16   CTE_free_protected_memory( pv );
17   void *pv_real = ((uint8_t *)pv) -
         PROT_ZONE_SIZE;
18   // call the real FreeRTOS vPortFree
         function
19   __real_vPortFree( pv_real );
20 }
```

Fig. 6.5 Wrappers for the real FreeRTOS *pvPortMalloc* and *vPortFree* memory management functions

Figure 6.5 shows our wrapper functions. The *pvPortMalloc* wrapper allocates a larger than requested memory block, by adding additional bytes (protected zones) before and after the requested memory block. These protected zones are registered in CTE (Line 10). CTE will monitor all load/store operations and trigger a simulation error in case of a write (or read) access inside of the zones. CTE generates trace conditions in case of a symbolic memory address. The *vPortFree* wrapper unregisters the two corresponding protected zones from CTE (Line 16), which does also check for double free and non-allocated blocks, and then calls the real *vPortFree* function of FreeRTOS.

Test Results We run CTE until we find the first error. We then repeatedly fix the error and re-run CTE, until no more error is found. We have set 4 h (14,400 s) as time limit to terminate the analysis. Table 6.2 presents the errors that we have found. All errors are related to buffer overflows of heap allocated memory and can therefore lead to serious security vulnerabilities. Besides a description, Table 6.2 shows additional relevant information similar to Table 6.1.

It can be observed that our concolic analysis, despite currently being without sophisticated state-of-the-art search heuristics, is already very effective in finding errors. Some of the detected errors require very specific inputs that are hard to find without formal methods. Note that these errors/vulnerabilities were present within the FreeRTOS TCP/IP stack for a long time already. At the time of writing, it appears that other researchers might also have discovered these errors [61]. These have been fixed only very recently in the newest FreeRTOS Version (v10.1.0). Despite being assigned CVEs (detailed list in [61]), neither the errors nor how they have been found are disclosed.

Table 6.2 Errors found in testing the FreeRTOS TCP/IP stack implementation—time in seconds

Error description	Time	stime	#paths	#queries	#instr
1: A malformed IP packet header length causes an integer overflow which leads to a *memmove* operation with a size close to UINT_MAX.	0.52	0.13	7	30	1,106,862
2: Multiple buffer overflows accessing (read) non-existing fields in the DNS and NBNS packet parser.	4.52	0.58	62	155	10,256,275
3: Buffer overflow (write) in the DNS reply generator, due to missing packet length checks, causing heap corruption.	4.94	0.50	90	207	14,946,642
4: Various buffer overflows (read), during TCP options checking, due to missing buffer length checks (in case of malformed packets).	10.90	2.55	108	346	20,148,331
5: Integer overflow in length calculation causes NBNS processing code to allocate a large reply buffer and fill it by reading beyond a much smaller input buffer.	380.92	33.00	6135	15,495	1,025,290,638
6: NBNS processing code allocates not enough memory to hold the complete reply message for some malformed UDP packet sizes, leading to a buffer overflow.	456.88	36.23	8055	20,171	1,343,648,874

6.1.4 Discussion and Future Work

CTE is already very effective for testing embedded binaries targeting RISC-V systems with peripherals. To further improve CTE we plan to:

- Add support for the 64 bit RISC-V ISA and additional ISA extensions (e.g. floating point instructions) to support a broader range of RISC-V systems.
- Incorporate further state-of-the-art symbolic exploration techniques to speed up the verification process. In particular, we plan to integrate sophisticated state space exploration heuristics, which are very important to speed up the bug finding process in very large state spaces. In addition, we want to integrate dynamic state merging techniques to alleviate the state space explosion problem.
- Integrate a RISC-V processor timing model into CTE to obtain accurate timing results for systems with peripherals and enable checking timing related properties alongside the concolic testing.
- Support interrupts with symbolic trigger times to enable efficient checking of bugs related to different interrupt orderings. A viable solution might be to support intervals for each interrupt that denotes the minimum and maximum delay for triggering that interrupt.
- Investigate using C++ peripheral SW models with a more comprehensive abstraction layer to avoid the current peripheral transformation step.

6.2 Verification of Embedded Binaries using Coverage-Guided Fuzzing

Recently, in the SW domain the automated SW testing technique *fuzzing* [154] has been shown very effective in test-case generation and became a standard in the SW development flow [152, 164]. These modern fuzzing techniques typically employ the so-called *mutation* based technique. It mutates randomly created data and is guided by code coverage. Notable representatives in this *Coverage-guided Fuzzing* (CGF) category are the LLVM-based *libFuzzer* [141] and *AFL* [9], which both have been shown very effective and revealed various new bugs that can lead to errors and security vulnerabilities [9, 141]. However, using CGF for checking embedded SW binaries is challenging, because it requires to analyze architecture specific SW binaries that extensively interact with HW peripherals.

In this section we propose a novel approach that leverages state-of-the-art CGF methods in combination with SystemC-based VPs for verification of embedded SW binaries. We call this combination *VP-CGF*. VP-CGF brings together the benefits of both worlds: CGF is a sophisticated and very effective method for test-case generation and VPs allow for a fast and accurate binary-level SW analysis and enable checking of complex HW/SW interactions during test-case execution. To guide the fuzzing process we combine the coverage from the embedded SW additionally with the coverage of the SystemC-based peripherals. Both coverage information are tracked in the VP alongside the test-case execution. Our experiments demonstrate the effectiveness of our approach in analyzing real-world RISC-V embedded SW binaries.

Several formal verification methods [6, 39, 41, 101, 110, 111, 157], primarily based on symbolic/concolic execution techniques, have been proposed for verification of embedded SW binaries. However, formal methods are still highly susceptible to state explosion.

Different random testing/fuzzing approaches targeting embedded systems have been reported: [125, 138] use random fuzzing of automobile ECUs with CAN packets. Alimi et al. [8] and Kamel and Lanet [117] use generational (model-based) and evolutionary (based on genetic algorithms) fuzzing techniques to analyze smart cards. Mulliner et al. [158] and van den Broek et al. [211] analyze the GSM implementation in mobile phones by sending random SMS messages. However, neither of these approaches leverages state-of-the-art CGF or VPs. An overview on challenges in fuzzing embedded systems is provided in [156].

In [118, 223] methods are presented to integrate real HW peripherals with the SW simulation to enable accurate analysis of embedded SW even when no peripheral models are available. We consider these approaches to be complementary to our approach.

Recently, QEMU has been combined with AFL and even been used for fuzzing the Linux kernel [174, 208]. This combination (QEMU-AFL) enables CGF-based analysis of embedded binaries by emulating them in QEMU and propagating the observed SW coverage at runtime from QEMU back to AFL. Another similar project combines AFL with the Unicorn CPU emulator [3]. While these approaches are arguably the closest to ours, as they also enable CGF of embedded binaries, they are primarily focused on emulating the CPU core and either do not consider peripherals at all (Unicorn-AFL) or only to a limited extent (QEMU-AFL). In addition, QEMU operates on a different abstraction level than SystemC-based VPs which can hinder accurate full platform simulations, and QEMU-AFL does not consider the peripheral coverage in the fuzzing process.

6.2.1 VP-based Coverage-Guided Fuzzing

In this section, we present our proposed approach *VP-CGF* for verification of embedded SW binaries using CGF with SystemC-based VPs. We start with an overview (Sect. 6.2.1.1). Then, in Sect. 6.2.1.2 we describe how we track code coverage information for the SystemC-based peripherals (in the VP) and the embedded SW binary (that is tested). Next (Sect. 6.2.1.3), we present an example embedded application that is verified with VP-CGF (to further illustrate the concept of VP-CGF, the real experiments are presented later in Sects. 6.2.2 and 6.2.3). Finally, in Sect. 6.2.1.4 we provide a discussion on encoding functional coverage information, tailored for embedded applications, to complement code coverage.

6.2.1.1 Overview

Figure 6.6 shows an overview of our approach VP-CGF for verification of embedded SW binaries. VP-CGF leverages two components, a CGF-based fuzzer (shown on the left side of Fig. 6.6) and a SystemC-based VP (shown on the right side of Fig. 6.6) that interoperate in a loop. Essentially, the fuzzer provides new inputs and the VP executes these inputs and returns coverage information to the fuzzer. VP and fuzzer are separate processes and communicate through sockets and files.[4] In the following we present more details on our approach.

Starting point is an embedded SW binary that represents the embedded SW application. In the first step the VP is started in *server-mode* and the embedded SW binary is loaded into the VPs memory. When started in *server-mode*, the VP will fully initiate (i.e. set up the SystemC simulation engine and instantiate all components) until the VP is ready for execution of the embedded SW binary. At this point the VP will stop and wait for commands from the fuzzer.

[4]Other interprocess communication mechanisms, like shared memory, can be used as well. Please note, running the fuzzer in the same process as the VP does not work because the SystemC simulation engine does neither provide a reset function nor can it be re-instantiated without re-starting the whole process. Hence, it would not be possible to run multiple fuzzer inputs starting from the initial state.

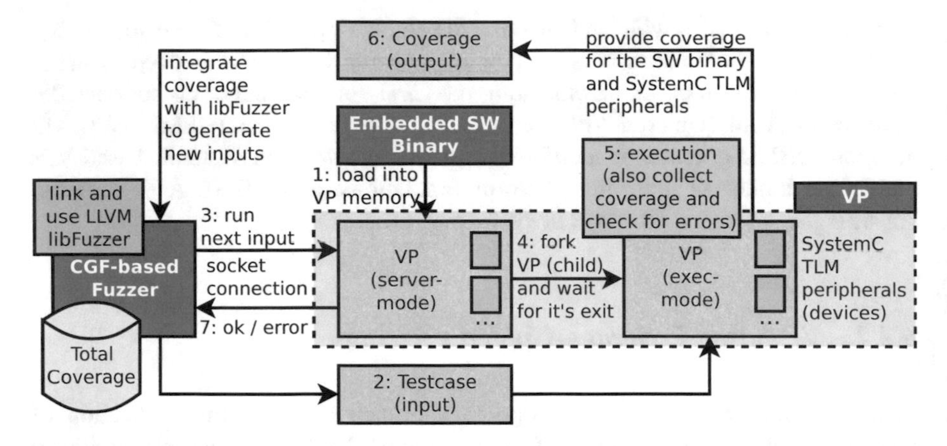

Fig. 6.6 Overview on VP-CGF: CGF with SystemC-based VPs for verification of embedded SW binaries

Next, we start the fuzzer that connects with the VP through a socket connection and starts the *fuzz-loop* (repeating Steps 2–7 in Fig. 6.6), which works as follows: The fuzzer begins by writing a new testcase into a (binary) file (Step 2, bottom of Fig. 6.6) and notifies the *server-mode* VP by sending a run command (Step 3). The VP will then use the *fork* system call to create a new identical VP child process in *exec-mode* (Step 4) to execute the embedded SW binary with the fuzzer-provided input (Step 5, right side of Fig. 6.6). Alongside the execution, the *exec-mode* VP collects coverage information for the embedded SW binary as well as the SystemC peripherals, and writes them into a coverage output file (Step 6, top of Fig. 6.6). The *server-mode* VP waits until the *exec-mode* VP finishes (i.e. the forked process exits) and notifies the fuzzer that the execution has finished and whether an error was detected during this execution (Step 7), by checking the return code of the *exec-mode* VP process. The fuzzer reads and integrates the coverage output file. In case new coverage has been obtained, the fuzzer collects the testcase. The *fuzz-loop* continues (repeating Steps 2–7) until an error is detected or a user-defined timeout is reached. Based on the collected testcases a coverage report can be generated by re-running the testcases on the VP (in *exec-mode*).

Please note, using the *fork* system call to spawn a new VP instance for each fuzzer input has been very important to obtain good performance results. We observed speed-ups of more than 20x compared to starting a new VP instance (i.e. new process) for each fuzzer input. The main reason for this performance impact is the SystemC simulation engine which requires significant time for an initial startup.

6.2.1.2 SW and Peripheral Coverage Collection in the VP

During execution the VP collects coverage from the embedded SW binary and from the SystemC peripherals. Currently, we focus on branch coverage information.

Peripheral Coverage We obtain branch coverage information for the SystemC peripherals by compiling them with Clang using the *-fsanitize-coverage=trace-pc-guard* option. With this option Clang instruments the peripherals to emit coverage information for branch instructions at runtime by calling special interface functions. We provide these interface functions in the VP to collect the coverage information and forward that information, through the coverage output file, to the fuzzer accordingly. The fuzzer integrates the coverage by forwarding it to libFuzzer, since libFuzzer also understands Clang instrumentation and hence provides the same interface functions. Please note, we only selectively instrument the SystemC peripherals with Clang and not the whole VP to avoid the (potentially) significant runtime overhead of instrumenting (in particular) the ISS of the VP and also avoid the communication overhead in transferring this additional coverage information (which is irrelevant for testing the embedded SW) to the fuzzer.

Embedded SW Coverage We obtain (branch) coverage information for the embedded SW binary by observing the executed instructions in the ISS of the VP. In particular, we have installed a hook in the ISS that intercepts every instruction execution and checks if it is a branch instruction. In case of a branch instruction we collect the address of the branch instruction itself (pre-pc) and the address of the instruction after the branch (post-pc, which depends on whether the branch was taken or not). The pair (pre-pc, post-pc) represents a control flow edge. We collect all observed edges in the Clang instrumentation format to keep it compatible with libFuzzer. This also allows us to store all coverage information (for the embedded SW binary and SystemC peripherals) in a single unified list. Essentially, Clang assigns each branch instruction a unique *guard* index. We mimic this behavior by mapping each different edge to an ascending index starting after the last *guard* index used by Clang for instrumenting the peripheral branches (Clang provides an interface function to obtain the number of instrumented branches).

6.2.1.3 Example Embedded Application Fuzzing

In this section we present an example embedded application that consists of an embedded SW accessing a sensor. The example illustrates how SW and peripherals access fuzzer-provided input and interact with each other. It also demonstrates basic SystemC modeling concepts. We start by presenting the SystemC implementation of the sensor (Sect. 6.2.1.3) and then show the embedded SW accessing this sensor (Sect. 6.2.1.3). As foundation for this example application, we used the sensor implementation from the open-source RISC-V VP that has been presented in Chap. 3.

SystemC TLM Sensor Peripheral in VP Figure 6.7 shows the sensor implementation in SystemC TLM. The sensor model has two 32 bit configuration registers *scaler* and *filter* (Lines 10–11) and a data frame of 8 bytes (Line 8) that represents the current sensor measurement and is periodically updated with new data (Lines 37–46). The scaler register setting defines the speed at which the sensor data frame is updated. The filter register setting selects the sensor internal post-processing that is applied on the data frame.

```
1  struct Sensor : public sc_core::sc_module {
2    tlm_utils::simple_target_socket<Sensor> tsock;
3    std::shared_ptr<InterruptController> IC;
4    sc_core::sc_event run_event; // SystemC event
5    // all peripherals can access the fuzzer-provided input through a shared pointer
6    std::shared_ptr<fuzz_input_if> fuzz_input;
7    // memory mapped data frame
8    std::array<uint8_t, 8> data_frame;
9    // memory mapped configuration registers
10   uint32_t scaler = 25;
11   uint32_t filter = 5;
12   std::unordered_map<uint64_t, uint32_t *> addr_to_reg;
13   enum {
14     SCALER_REG_ADDR = 0x80, // local addresses
15     FILTER_REG_ADDR = 0x84, // in the sensor
16   };
17
18   SC_HAS_PROCESS(Sensor); // allow SystemC processes
19   Sensor(sc_core::sc_module_name) {
20     // register a handler for TLM transactions
21     tsock.register_b_transport(this, &Sensor::transport);
22     // declare the run function to be a SystemC thread
23     SC_THREAD(run);
24     // mapping to conveniently access registers
25     addr_to_reg = {
26       {SCALER_REG_ADDR, &scaler},
27       {FILTER_REG_ADDR, &filter},
28     };
29   }
30
31   void run() {
32     while (true) {
33       // run periodically based on the scaler register
34       run_event.notify(sc_core::sc_time(scaler, sc_core::SC_MS));
35       sc_core::wait(run_event);
36       // fill data frame
37       for (auto &n : data_frame) {
38         // consume next byte from fuzzer-provided input (stop simulation if empty) which represents the environment measurements
39         uint8_t x = fuzz_input->next();
40         // apply sensor internal post-processing of the input based on the filter register setting
41         if (filter < 3) {
42           n = std::max((uint8_t)245, x);
43         } else {
44           n = x % 100;
45         }
46       }
47       // trigger interrupt to notify the SW
48       IC->trigger_interrupt(2 /*IRQ number*/);
49     }
50   }
51 };
```

Fig. 6.7 Example sensor peripheral implemented in SystemC TLM

In this example we use fuzzer-provided input to update the sensor data frame. Hence, the fuzzer-provided input represents the sensor measurements obtained from the environment. The fuzzer-provided input is shared among all (SystemC) peripherals, using a shared *fuzz_input* pointer (Line 6). Calling *fuzz_input->next()* (Line 39) consumes and returns the next byte from the fuzzer-provided input. This raw data byte is then post-processed based on the filter register setting (Lines 41–45). The simulation is stopped once all fuzzer input has been consumed (in any peripheral). The data frame update happens in the run thread (the run function is registered as SystemC thread inside the constructor in Line 23). Based on the scaler register value the run thread is periodically triggered (Line 34) by calling the

```
52  // SW load/store instructions
        directed to the sensor address
        range are routed by the VP bus
        to this function through TLM
        sockets
53  void transport(
        tlm::tlm_generic_payload &trans,
        sc_core::sc_time &delay) {
54    auto addr = trans.get_address();
55    auto cmd = trans.get_command();
56    auto len = trans.get_data_length();
57    auto ptr = trans.get_data_ptr();
58
59    if (addr <= 7) { // access data
          frame
60      assert(cmd ==
            tlm::TLM_READ_COMMAND);
61      assert((addr + len) <=
            data_frame.size());
62      // return last generated random
            data at requested address
63      memcpy(ptr, &data_frame[addr],
            len);
64    } else {
65      assert(len == 4); // NOTE: only
            allow to read/write whole
            register
66      auto it = addr_to_reg.find(addr);
67      assert (it != addr_to_reg.end());
            // access to non-mapped
            address
68      // trigger pre read/write actions
69      if ((cmd ==
            tlm::TLM_WRITE_COMMAND) &&
            (addr == SCALER_REG_ADDR)) {
70        uint32_t value = *((uint32_t
              *)ptr);
71        if (value < 1 || value > 100)
72          return; // ignore invalid
                values
73      }
74
75      // actual register read/write
76      if (cmd == tlm::TLM_READ_COMMAND)
          {
77        *((uint32_t *)ptr) = *it->second;
78      } else if (cmd ==
            tlm::TLM_WRITE_COMMAND) {
79        *it->second = *((uint32_t *)ptr);
80      } else {
81        assert(false && "unsupported tlm
              command for sensor access");
82      }
83
84      // trigger post read/write actions
85      if ((cmd ==
            tlm::TLM_WRITE_COMMAND) &&
            (addr == SCALER_REG_ADDR)) {
86        // re-schedule the run event
              with the new scaler value
87        run_event.cancel();
88        run_event.notify(sc_core::sc_time(
              scaler, sc_core::SC_MS));
89      }
90    }
91  }
```

Fig. 6.8 Implementation of the *transport* function for the sensor peripheral example shown in Fig. 6.7

notify function on the internal SystemC synchronization event. After every update an interrupt is triggered, which will propagate through the interrupt controller (IC, Line 3) to the CPU core up to the interrupt handler in the application SW.

The SW can access the sensor data frame and configuration registers through memory mapped I/O. Each SW load/store instruction is translated into a TLM transaction and routed by the VPs bus system accordingly. Transactions with addresses matching the sensor address range will be delivered to the transport function of the sensor that is shown in Fig. 6.8.[5]

[5]This routing also involves a *global-to-local* address translation (i.e. the sensor is mapped to address 0x50000000 and the SW accesses the global address 0x50000080, then the bus system translates the address to the sensor local address 0×80 before delivering the transaction to the sensor).

```
1  // memory mapped register/inputs of the peripherals
2  volatile char *SENSOR_INPUT_ADDR = (char*)0x50000000;
3  volatile uint32_t *SENSOR_SCALER_REG_ADDR = (uint32_t*)0x50000080;
4  volatile uint32_t *SENSOR_FILTER_REG_ADDR = (uint32_t*)0x50000084;
5  volatile char *FUZZER_INPUT_ADDR = (char*)0xC0000000;
6
7  uint8_t fuzz_u8() {
8    // consume the next byte from the fuzzer-provided input, corresponds to "fuzz_input->next()" in Fig. 6.7 (Line 42 and Line 44)
9    return *FUZZER_INPUT_ADDR;
10 }
11 // read four bytes from the fuzzer-provided input
12 uint32_t fuzz_u32() {
13   return (fuzz_u8() << 24) | (fuzz_u8() << 16) | (fuzz_u8() << 8) | fuzz_u8();
14 }
15
16 _Bool has_sensor_data = 0;
17 void sensor_irq_handler() {
18   has_sensor_data = 1;
19 }
20 void process_sensor_data() {
21   // wait for sensor interrupt
22   while (!has_sensor_data)
23     asm volatile ("WFI"); // Wait For (any) Interrupt
24   has_sensor_data = 0;
25
26   // read and validate sensor data
27   unsigned sum = 0;
28   for (int i=0; i<8; ++i) {
29     uint8_t n = *(SENSOR_INPUT_ADDR + i);
30     sum += n;
31   }
32   assert (sum < 8*250);  // check that sensor data stays within a valid maximum range
33 }
34
35 void main() {
36   register_interrupt_handler(2 /*IRQ_NUMBER*/, sensor_irq_handler);
37   // configure the sensor based on fuzzer input
38   *SENSOR_SCALER_REG_ADDR = fuzz_u32();
39   *SENSOR_FILTER_REG_ADDR = fuzz_u32();
40   process_sensor_data();
41 }
```

Fig. 6.9 Example embedded SW accessing the (SystemC TLM) sensor peripheral shown in Fig. 6.7

Based on the address and operation mode (Lines 54–55), as stored in the transaction object (generic payload), the action is selected. It will either read (part of) the data frame (Line 63) or read/write one of the configuration registers (Lines 76–82). In case of a register access a pre-read/write (Lines 69–73) validation and post-read/write action (Lines 85–89) can be defined as necessary. In this example, the sensor will ignore invalid scaler values (Lines 71–72) and reset the data generation thread on a scaler update (Lines 87–88).

Embedded SW Figure 6.9 shows an example embedded SW that accesses the sensor. The SW starts by installing an interrupt handler (Line 36) and configures the sensor *filter* (Line 39) and *scaler* setting (Line 38) by writing to designated memory mapped registers. Then, the SW waits for a sensor interrupt (Lines 22–23), which denotes that new sensor data is available. Finally, the SW fetches the sensor data (Lines 27–31) and validates it (Line 32).

Both sensor configuration registers are initialized with fuzzer-provided input (via the *fuzz_u32* function defined in Line 12). The SW can access the fuzzer-provided input through a special peripheral that maps the fuzzer input into the memory. Reading a byte from this memory location (Line 9) will consume and return a byte from the fuzzer-provided input (it corresponds to *fuzz_input - >next()* as shown for the sensor in Fig. 6.7 in Line 39). The (SystemC) simulation is stopped by the (SystemC) peripheral in case all fuzzer-provided input has been consumed.

The assertion in the SW in Line 32, which checks that the sensor data stays within a valid range, can be violated. Please note, it is possible to maximize the SW code coverage but still miss the assertion violation because the occurrence of the violation depends on the filter setting of the sensor (i.e. the value that the SW writes in Line 39 and the sensor then accesses in Fig. 6.7 in Line 41). This example demonstrates that it is important to also consider code coverage from peripherals during testing of embedded SW to detect bugs that depend on HW/SW interactions. With our VP-CGF approach we were able to detect this bug within a few seconds.

6.2.1.4 Discussion: Encoding Functional Coverage for Embedded Systems

The example in Sect. 6.2.1.3 demonstrated that it is important to also consider the peripherals during testing of embedded SW due to (extensive) HW/SW interactions. In particular accesses to configuration registers are important, since different values can cause (significantly) different behavior of the peripheral. Thus, it is also important to consider different configuration register values in the coverage metric. This can already happen implicitly by tracking the branch coverage if the control flow depends on the register values, as shown in the sensor example (Fig. 6.7 in Line 41, where the control flow depends on the filter register value). An alternative (i.e. the control flow does not already depend on the register values) is to add artificial empty branches, e.g. *if (config_reg == 1) __asm__ volatile ("");*, to explicitly encode different register values as code coverage. The *__asm__ volatile* instruction ensures that the empty branch is not omitted by the compiler. Configuration registers with a small set of different values can be fully enumerated (it is possible to define macros that help in generation of these branches). However, in case of a large set of possible values, expert knowledge is required to define suitable equivalence classes for different register values.

Another important property that is very specific to embedded systems, and hence need to be considered during testing as well, is the execution order of peripheral functionality (i.e. the HW) which in SystemC is encoded as threads (processes in general). Depending on the thread scheduling order, interrupts can occur at different times, and SW functions that interact with HW components can behave differently (e.g. a data transfer application that blocks either during reading or writing depending on the processing speed of the receive and transmit UARTs). Considering different thread scheduling orders can expose additional HW/SW interaction related bugs. The sensor example already demonstrated that fuzzer input can be used to configure the thread execution period (by setting the scaler register of the sensor). Beside encoding the execution period as coverage, it is also possible to encode the relation of the execution periods between different threads. For example for two threads t_1 and t_2 consider the three cases: period(t_1) < period(t_2), period(t_1) = period(t_2), period(t_1) > period(t_2). These three cases correspond to a setting where t_1 executes faster, equally fast and slower than t_2, respectively.

In the following we present our experiments, using RISC-V embedded SW binaries as examples.

6.2.2 *Experiment 1: Testing Embedded Applications*

We have implemented our proposed VP-CGF approach for verification of embedded binaries using LLVM libFuzzer (Sect. 2.3) and the open-source RISC-V based VP (Chap. 3) as a case study. We evaluate VP-CGF in two steps: First, in this section, we present results on testing two embedded RISC-V applications with VP-CGF. Then, in the next section, we show results on applying VP-CGF to test the Zephyr OS network IP stack. All experiments are evaluated on a Linux machine with Ubuntu 16.04.1 and an Intel processor with 2.5 GHz.

In the following we start with an overview of the results for the first set of experiments and then present more details for each of the two embedded applications.

6.2.2.1 Results Overview

Table 6.3 gives the results. It shows 11 different metrics (one in each row) for both embedded applications. In particular, Table 6.3 shows the LoC (*Lines of Code*) of the embedded SW in C (Row 1) and RISC-V ASM (Row 2) as well as the LoC of the relevant SystemC peripherals (Row 3) in the VP. Next, Table 6.3 shows the time in second until VP-CGF finds the bug (Row 4), how many testcases have been executed in total (Row 5) as well as the average (Row 6) and maximum (Row 7) number of test-case executions per second. Row 8 shows the final number of collected testcases (i.e. testcases that increase the coverage, until the bug is found) and Rows 9 and 10 show the line coverage of the SW and SystemC peripherals that we obtain by re-running the testcases of Row 8. Finally, Row 11 shows the time in seconds to find the bug by generating testcases purely randomly (without any coverage feedback) as comparison to VP-CGF.

6.2.2.2 Application 1: Data Transfer

The first application that we consider reads a data stream from an input device, processes the data stream and writes the processed stream back to an output device. We use a UART in this application to act as input (RX) and output (TX) device, respectively. The data is stored in an internal ring buffer during the copy process in

Table 6.3 Experiment results on applying our VP-CGF approach for testing embedded applications

Metric	Application	
	(1) data transfer	(2) fan control
1: LoC SW C	142	116
2: LoC SW RISC-V ASM	435	371
3: LoC Peripherals SystemC	331	262
4: VP-CGF Time to Bug (s)	180	20
5: VP-CGF Exec Total	22,189	6400
6: VP-CGF Exec/Sec Average	123	320
7: VP-CGF Exec/Sec Maximum	651	640
8: VP-CGF Number Final Tests	98	46
9: VP-CGF Coverage SW	98.18%	97.5%
10: VP-CGF Coverage Peripherals	94.79%	96.35%
11: VP-Random Time to Bug (s)	T.O. (>3600)	3212

the SW. The application starts initializing the system, i.e. setting up peripherals by writing to their configuration registers and registering interrupt handlers, and then enters a WFI (*Wait For Interrupt*) loop. On each UART interrupt (i.e. the UART has new incoming data to receive or is able to transmit new outgoing data) the application logic is triggered.

The application logic consists of two parts: (1) First, it transfers data in a loop from the input UART to the ring buffer. Figure 6.10 shows an overview. The loop stops when the UART receive queue is empty (Line 7) or the ring buffer has not sufficient free space to store new data (Line 2). The functions *is_full* and *get_free_space* return whether the ring buffer is full and number of free entries, respectively. Two operation modes are available: *raw* and *escape* (as indicated by the Boolean *raw_mode* variable). In *raw* mode the data is simply passed on unprocessed (Line 13). In *escape* mode the special character `0x7f` is escaped by *ESC_CHAR*, i.e. *ESC_CHAR* is inserted into the stream right before `0x7f` (Lines 10–11). Whenever the `0xff` byte is observed in the input stream the operation mode is changed from *raw* to *escape* mode and vice versa (Line 15). (2) In the second part of the application logic, the ring buffer content is transferred to the output UART. The transfer loop is similar to Fig. 6.10. It stops when the transmit queue of the output UART is full or the ring buffer is empty.

Test Setup On the SW side we initialize the UART RX and TX interrupt *watermark-level* with fuzzer-provided input. The *watermark-level* configures when interrupts are triggered by the UART depending on how many elements are currently stored in the RX and TX queue, respectively. We added coverage points (on the VP side, when the configuration is received in the SystemC peripheral) to distinguish between the different possible *watermark-levels* (0..7 for RX and 0..7 for TX). On the VP side we fill the RX queue of the UART with fuzzer-provided input and check that the data arrives again in the same order in the TX queue of the UART and is

```
1  while (1) {
2    if ((raw_mode && is_full(&buf)) || (!raw_mode && (get_free_space(&buf) < 2)))
3      break; // ring buffer has not enough free space
4    uint32_t rx = *UART_RX;
5    if (rx & (1 << 31)) {
6      // UART receive queue is empty
7      break;
8    }
9    uint8_t data = rx & 0xff;
10   if (!raw_mode && (data == 0x7f)) {
11     write(&buf, ESC_CHAR);
12   }
13   write(&buf, data);
14   if (data == 0xff)
15     raw_mode = !raw_mode;
16 }
```

Fig. 6.10 SW loop to read and process characters from the input (RX) UART

correctly escaped (after being processed by the SW). Therefore, we use SystemC threads with periodic delays (RX-delay and TX-delay, respectively) to (1) add new (fuzzer-provided) data to the RX queue of the input UART (RX-thread), and (2) consume data from the TX queue of the output UART (TX-thread). Both cases can trigger an interrupt, depending on the interrupt *watermark-level* configuration of the UART, that will in turn notify the SW application.

Please note, that the SW application is non-terminating, because it waits for new interrupts indefinitely. Therefore, we simply stop the SW execution in the VP when the whole fuzzer input has been consumed.

We also use the fuzzer-provided input to configure the RX- and TX-delays, respectively. This allows to check for errors related to the execution order of the RX-/TX-threads. We have added coverage points to distinguish the three cases: RX-delay $<$ TX-delay, RX-delay $=$ TX-delay, RX-delay $>$ TX-delay. These cases describe settings where data can be faster/equally fast/slower received than transmitted.

The RX (receive) and TX (transmit) queues of the UART have each a size of 8. The SW ring buffer has a size of 16.

Results We found a bug with VP-CGF in 180 s. The bug results in data being overwritten in the ring buffer before it is written back to the (TX) UART. The reason for the bug is an incorrect length calculation in the ring buffer. In particular, the *get_free_space* function returns one more element than available if the buffer *read* pointer is before the buffer *write* pointer (*read* and *write* denote the next element to be read/written, respectively. They are updated after each access accordingly and wrap around the ring buffer when reaching the end). The bug is triggered if *read* points to the first element and *write* to the second last element of the ring buffer while the SW Boolean variable *raw_mode* is *false* and at the same time the data byte `0x7f` is received from the UART. In this case the SW will pass through the check in Line 2 without hitting the break (Line 3) and then the ring buffer will overflow by writing two elements (Lines 11 and 13) to it.

It can be observed that VP-CGF is much more efficient (at least factor 20.0x faster in finding the bug) than pure random test generation, which received a timeout after 3600 s (1 h). We also obtained very high coverage values for both the SW (98%) and the SystemC-based peripherals (94%). Careful examination of the coverage results reveals that the remaining 6% peripheral coverage is indeed unreachable in combination with this SW. It corresponds to access of additional configuration registers (in the UART) and interrupt priorities (interrupt controller), which are not used by this SW. The remaining missing 2% SW coverage can be closed by re-running VP-CGF on the fixed SW.

6.2.2.3 Application 2: Fan Control

The second application that we consider controls the system fan speed based on temperature values obtained from a temperature sensor. Therefore, the sensor triggers periodic interrupts after each time interval to notify application that a new sensor data frame is available. Each data frame contains the temperature values measured during the last time frame. The sensor provides a filter register that controls the range of measured temperature values. On each sensor interrupt, the application first copies the sensor data frame into an internal buffer and then computes the average temperature value for the last time frame (i.e. for the elements of the buffer). Based on the current average temperature value together with the last observed average temperature value the fan speed setting is controlled (level=0 off to level=5 high speed). The fan speed is immediately set to high speed in case of a high temperature value but requires low temperature values for two time frames to reduce its speed. Technically, the mapping from temperature value to fan speed level is implemented by normalizing the temperature value to an array index and accessing a pre-configured buffer.

Test Setup On the SW side we initialize the sensor filter register value with fuzzer-provided input. On the VP side we fill the sensor data frame with fuzzer-provided input. Similar to the first application, this application is also non-terminating, and hence we also stop the execution once the whole fuzzer input has been consumed. We added assertions in the fan control peripheral to check that a high fan speed is selected in case a high temperature value is measured. In addition, we have added assertions to check for buffer overflows in the SW, i.e. by adding bounds checks before every array access.[6]

[6]Please note, in general bounds check assertions can be added automatically by using sanitizers to instrument the source code. However, for embedded systems sanitizer support is often still missing (e.g. error detection sanitizers are missing for the RISC-V architecture). Nonetheless, it is possible to detect buffer overflows of heap allocated memory with comparatively little effort by performing a dynamic instrumentation as was done for testing the Zephyr network IP stack (see Sect. 6.2.3.3).

Results We found a buffer overflow with VP-CGF in 20 s. The reason for this error is a bug in the temperature normalization procedure that restricts all temperature values above a certain threshold to stay below that threshold. However, the case where the (average) temperature value equals the threshold is missed and hence this particular temperature value is not below the threshold, which in turn results in a computation of an out-of-bounds buffer index. To trigger the buffer overflow the temperature sensor (hence the fuzzer in this setting) has to provide a data frame such that the average temperature value of this frame is equal to the threshold. Please note that the data frame also depends on the filter register value in the sensor which is also provided by the fuzzer.

In this experiment, it can also be clearly observed that CGF is much more effective than pure random generation (a factor of 160x faster in finding the bug). Similar to the first experiment we also obtained very high line coverage results in both the SW (97%) as well as the SystemC-based peripherals (96%). In fact, the remaining 4% coverage in the peripherals are indeed unreachable code in combination with this SW. It corresponds to configuring interrupt priorities in the interrupt controller, which is not used by the SW. The remaining missing 3% SW coverage can be closed by re-running the fuzzer on the fixed SW.

6.2.3 Experiment 2: Testing the Zephyr IP Stack

In the second experiment we apply our approach VP-CGF to test the network IP stack of the Zephyr OS in combination with the RISC-V port of the Zephyr kernel (Zephyr version 1.14.99). This experiment demonstrates that VP-CGF is applicable to analyze larger real-world embedded SW. We test for generic execution errors (e.g. SW assertion violation and access of unmapped memory) and buffer overflows in particular. We have integrated a custom mutator into the fuzzing process to further guide it. We start by explaining our test setup (Sect. 6.2.3.1), then present our custom mutator (Sect. 6.2.3.2) and method for buffer overflow (Sect. 6.2.3.3) detection at runtime. Finally, we present our results (Sect. 6.2.3.4).

6.2.3.1 Test Setup

IP packets are processed by the platform independent IP-thread in Zephyr (threads and synchronization mechanisms are provided by the Zephyr kernel). Packets are delivered to the IP-thread through a shared queue by a lower-level network driver.

In this experiment we use the SLIP (*Serial Line Internet Protocol*) network driver of Zephyr. SLIP allows to send and receive IP packets by serializing and deserializing the data. This in turn allows sending and receiving IP packets (and hence use higher-level protocols like UDP, TCP, etc. as well) through a serial driver in combination with a UART device.

The VP provides a UART implementation that matches the SiFive specification [195] for which a serial driver is available in Zephyr, hence we can directly configure the SLIP network driver to use that UART.

In this experiment, we treat the whole fuzzer-provided input as a single IP packet. We wrap the IP packet as SLIP packet and pass it to the UART (in the VP). This will in turn trigger a UART interrupt which notifies the serial driver and finally the SLIP driver. The SLIP driver extracts the IP packet and delivers it to the IP-thread of Zephyr.

In the (Zephyr) SW main function we create and bind an UDP and TCP socket using the IPv4 as well as IPv6 protocol. This ensures that the Zephyr IP-thread does not drop IP packets prematurely because no matching socket is bound. Please note, that the Zephyr IP-thread is non-terminating because it waits for new IP packets indefinitely. Therefore, we have added a switch in the IP-thread to stop execution after one IP packet has been processed.

6.2.3.2 Custom IP Packet Mutation

LLVM libFuzzer provides a designated interface function to easily integrate custom mutators into the fuzzing process. We added such a custom mutator and configured it to be called with the same probability as the existing libFuzzer mutators. Our custom mutator simply overwrites the fuzzer-provided data with a small pre-defined UDP packet. Our mutator starts overwriting the data from the beginning and stops when either the whole packet has written or the end of the fuzzer-provided data is reached, hence only injecting a prefix of the packet in this case. The injected UDP packet is completely valid and will fully propagate through the IP stack to the UDP socket and reach the Zephyr application when send independently. Our mutator essentially introduces valid IP and UDP header options into the fuzzing process to enable a more comprehensive evaluation.

6.2.3.3 Heap Buffer Overflow Detection

Zephyr provides several different functions to manage memory. These include the functions *k_malloc* and *k_free*, which correspond to the well-known *malloc* and *free* functions of the C library, as well as specialized functions optimized for allocation of fixed size memory blocks. We have instrumented the memory management functions of Zephyr, by wrapping them, to keep track of allocated memory blocks and their bounds. The instrumentation is conceptually similar to the instrumentation that we performed for FreeRTOS (see Sect. 6.1.3.2).

As an example, Fig. 6.11 shows wrappers for *k_malloc* and *k_free*. The *k_malloc* wrapper calls the real *k_malloc* function of Zephyr to allocate a larger than requested block (Lines 4–7), by adding extra space (*ALLOC_MARGIN*) before and after the requested block. The extra space is registered in the VP (Line 15), alongside the allocated block, and treated as protected memory. The VP monitors all memory

```
1  #define ALLOC_MARGIN 128 // in bytes
2
3  void *k_malloc(size_t size) {
4    size_t alloc_size = size +
         2*ALLOC_MARGIN;
5
6    // call the real k_malloc of Zephyr
7    void *p = real_k_malloc( alloc_size
         );
8
9    if (p == NULL)
10     return NULL;
11
12   void *block = ((uint8_t*)p) +
         ALLOC_MARGIN;
13
14   // notify the VP about the
         allocation
15   VP_k_malloc( block, size,
         ALLOC_MARGIN );
16
17   return block;
18 }
19
20 void k_free(void *block) {
21   // notify the VP about the free
22   VP_k_free( block );
23
24   void *p = ((uint8_t*)block) -
         ALLOC_MARGIN;
25
26   // call the real k_free function of
         Zephyr
27   real_k_free( p );
28 }
```

Fig. 6.11 Wrapper for dynamic memory management functions to keep track of the currently (dynamically) allocated blocks and check for buffer overflows at runtime

access operations (read and write) during execution and triggers an error in case the address falls in a protected memory region. The *k_free* wrapper notifies the VP about the *k_free* call (Line 22), to unregister the protected memory margins and check for double free as well as non-allocated block errors, and finally calls the real *k_free* function of Zephyr (Line 27).

6.2.3.4 Results

We have obtained 82.44% line coverage in the IPv4 and IPv6 processing stack of Zephyr by running VP-CGF for 1 h. The resulting testset contains 575 testcases. VP-CGF executed 276,395 testcases in total with an average and maximum of 76 and 165 testcases per second, respectively. The analyzed RISC-V binary has 46,105 lines of ASM code.

To also evaluate the effectiveness of VP-CGF in detecting (buffer overflow) related bugs, we manually added a bug into the Zephyr IP stack. The bug corresponds to a typical overflow caused by incomplete buffer length checks which may result in severe security vulnerabilities. Similar bugs had been present in recent versions of FreeRTOS. In order to trigger that bug, the fuzzer has to generate an almost valid IPv4 header (otherwise the packet will be dropped in the IP stack before having a chance to trigger the bug) but with a specific malformed header length. VP-CGF detected this bug in 229 s and executed a total of 20,535 testcases.

6.2.4 Discussion and Future Work

In this section we proposed VP-CGF, an approach that leverages state-of-the-art CGF methods in combination with SystemC-based VPs for verification of embedded SW binaries. In addition to the coverage of the embedded SW, we also integrate the coverage of the (SystemC-based) peripherals to further guide the fuzzing process. Our experiments showed very promising results and demonstrate the effectiveness of VP-CGF in analyzing real-world embedded SW binaries. To further improve VP-CGF, for future work we plan to:

- Support stronger coverage metrics. In particular, we want to add support for Clangs data flow related coverage sanitizers (i.e. *CMP* instruction sanitizer) and investigate cross coverage metrics between the SW and peripherals, instead of considering them independently.
- Evaluate the impact of parallelization on the fuzzing process. In general our current architecture allows to run multiple VP-CGF sessions (each session has a separate VP and fuzzer instance) in parallel, since libFuzzer supports (safe multi-instance) coverage synchronization using the OS file system.
- Integrate additional dynamic error checking mechanisms into the VP to detect additional error classes at runtime. Error detection is complementary to test-case generation. We would also immediately benefit from error detection sanitizers once they have been ported for the RISC-V platform (at least when the source code for the embedded SW is available).
- Implement architectural improvements targeting performance optimizations, e.g. use a shared memory for a direct and faster communication between fuzzer and VP.
- Consider verification of multi-threaded programs and interrupts by leveraging advanced symbolic verification techniques, such as [87, 97] for concurrent C programs using POSIX threads, at the VP level.

6.3 Summary

This chapter presented two novel VP-based SW verification approaches to improve the VP-based design flow. In contrast to the standard VP-based design flow, where mainly simulation-based methods are employed that either require significant manual effort for test-case generation and refinement or suffer from limited coverage results, the proposed approaches enable an automated and comprehensive test-case generation.

The first approach leverages concolic testing for verification of embedded SW binaries targeting RISC-V systems with peripherals. It tightly integrates the *Concolic Testing Engine* (CTE) with the architecture specific *Instruction Set Simulator* (ISS) of the VP to achieve a high simulation performance at binary level. In addition a designated CTE-interface is provided to integrate (SystemC-based) peripherals into the concolic testing by means of SW models with comparatively little effort. The approach has been effective in finding several buffer overflow related security vulnerabilities in the FreeRTOS TCP/IP stack which demonstrates its applicability to real-world embedded applications.

The second approach leverages state-of-the-art CGF methods in combination with SystemC-based VPs for verification of embedded SW binaries. This combination enables a scalable and efficient test-case generation by employing CGF techniques and at the same time allows a fast and accurate binary-level SW analysis as well as checking of complex HW/SW interactions by using VPs. To further guide the fuzzing process the coverage from the embedded SW is combined with the coverage of the SystemC-based peripherals. The experiments, using RISC-V embedded SW binaries as examples, demonstrated the effectiveness of the proposed approach in maximizing the coverage and analyzing real-world embedded applications.

Since both approaches are complementary, we plan to investigate a combination for future work. The most basic combination would first run CGF and then switch to concolic testing once the coverage results stop to improve. The reason is that CGF enables an efficient and scalable exploration of a broad range of behavior while concolic testing leverages formal methods to pin-point specific paths in order to close the remaining coverage depths.

Chapter 7
Validation of Firmware-Based Power Management using Virtual Prototypes

Modern *System-on-Chips* (SoCs) must satisfy stringent requirements on power consumption and performance. With a continuously fast increase in number of implemented functionalities as well as in their complexity, meeting these requirements has become one of the major challenges in embedded system design. This new challenge demands a major shift in the design flow where power optimization/-management is no longer an afterthought. There is an industry-wide consensus that waiting for the availability of RTL is not feasible anymore, because once the RTL is written, power saving opportunities have already been greatly cut off [11]. As both Software (SW) and Hardware (HW) have a significant impact on the overall power consumption, early design steps at the system level, in particular HW/SW co-design, should take power into consideration.

On the other hand, the emergence of *Virtual Prototypes* (VPs) at the abstraction of *Electronic System Level* (ESL) has played a major role in modernizing the SoC design and verification flow. The much earlier availability as well as the significantly faster simulation speed in comparison to RTL is among the main benefits of SystemC-based VPs. These enable functional validation and verification as well as SW development very early in the design flow. Building on this success story, extending VPs to be power-aware to enable early power analysis is a very promising direction. Admittedly, RTL is the first stage where enough details are present to provide reasonably accurate power numbers, however, ESL power modeling and estimation techniques are rapidly getting better (see, e.g., [75, 162, 188]).

At the system level, the focus is not on low-level techniques such as power gating or dynamic voltage and frequency scaling but rather on fundamental design decisions that have a large impact on the power consumption, e.g. low-power architectures or *power management strategies*. The latter can contribute a great deal to the overall power saving by putting unused components into low-power states and waking them up properly in an intelligent manner. In most modern SoCs, the global power management strategy is implemented in firmware (FW) with the main advantages being the relative ease to develop and the flexibility in

V. Herdt et al., *Enhanced Virtual Prototyping*,
https://doi.org/10.1007/978-3-030-54828-5_7

reconfiguring the strategy for different target applications. The recent advances in ESL power modeling and estimation enable to execute a particular SW application in FW/VP co-simulation and check whether its power budget and performance requirement are met. However, there are still a number of shortcomings with this basic simulation-based approach. First, production-level SW is not yet available in early design stages. Second, simulating a full SW stack can still be very time consuming, even at the speed of VPs. Third, a SW application is executed under some predetermined workloads (i.e. application and environment inputs). These workloads might very possibly miss rare corner cases where the power budget is exceeded or the performance constraint is violated.

To address these shortcomings, this book proposes a novel VP-based approach to assess the power-versus-performance trade-off of FW-based power management. The approach is presented in Sect. 7.1. Instead of executing real SW applications, the approach makes use of system-level workload scenarios. The main novelty of the approach is the modeling of workload scenarios based on *constrained random* (CR) techniques [222] that are very successful in the area of SoC/HW functional validation and verification. Each workload scenario corresponds to a system-level use case with a specific power consumption profile and is described by a set of constraints. The constraints define the set of legal concrete workloads that are conform to the intended use case. The constraint-based description enables automated generation of a large number of different workloads within the scenario, hence reducing the risk of missing a corner case. As there is no freely available VP with power modeling and estimation, as a case study the open-source LEON3-based VP SoCRocket [189] has been extended with power management features and a FW-based dynamic power management strategy has been implemented. The proposed approach is, however, not limited to a particular VP or power modeling technique. The approach has been published in [91, 99].

Although the constraint-based workload description enables automated generation of a large number of different testcases (corresponding to SW workloads), hence reducing the risk of missing a corner case, a coverage metric to objectively measure the quality aspect as well as to guide the generation of scenarios is still an important missing piece. At the very least, it is mandatory that all power states of each component are comprehensively exercised by the generated testcases. Recent experience from the industry [196] makes a case for using stronger metrics. The book argues that the power states from different components or power domains are not necessarily independent. This also applies to the context of FW-based PM, since this global management scheme can change the power state of several components simultaneously according to the implemented strategy. Therefore, an appropriate coverage metric must account for all possible combinations of these interdependent power states. The cross coverage of power states is such a metric.

Section 7.2 proposes a novel coverage-driven validation approach for FW-based PM. The main contribution is a feedback-directed workload generation algorithm that generates testcases in an automated manner in order to maximize the cross coverage of power states. The approach works in two phases: first a bootstrap phase is performed to obtain preliminary coverage information based on randomly generated testcases and then a coverage-loop phase to close the remaining coverage gaps. The coverage-loop works in (two) different generation modes and integrates a refinement loop to guide the test-case generation process. The applicability and efficiency of the proposed approach is demonstrated using the previously extended open-source SoCRocket VP [189] and four different PM strategies implemented in FW. The approach has been published in [98].

7.1 A Constrained Random Approach for Workload Generation

ESL power estimation provides the basic technique for comparing different architectural/implementation options by providing power estimates using simulation. It has been intensely investigated in academia and industry.

In academia, approaches using cycle-accurate architectural simulators (e.g. [221]), power models per functional unit (e.g. [131, 177]), and ESL design extensions by power models (e.g. [162, 188]) have been proposed.

As a next step methods which allow for including power management concepts (e.g. power states and power domains) have been developed. The PwARCH framework has been introduced in [150]. This framework follows the UPF principles and allows to add a power architecture to a SystemC TLM model such that different power design alternatives can be explored. A similar approach has been presented in [120] but it is based on meta-modeling techniques.

A complete HW/SW co-design exploration methodology w.r.t. power has been introduced in [75]. The authors of [217] proposed an exploration approach targeting power domain partitioning at ESL. A design space exploration approach for power-efficient distributed embedded applications has been presented in [187].

Among the commercial tools for ESL power estimation are for instance Virtualizer from Synopsys or Vista from Mentor Graphics.

However, while these solutions (both academic and commercial) finally enable the comparison of power consumption for different design alternatives, they assume that appropriate workloads are already provided (mostly in form of some existing SW benchmarks). If the provided workloads are not representative enough, especially w.r.t. intended system-level use cases, the comparison results might be misleading. The approach proposed in this book targets this issue by providing means to specify abstract workload scenarios and enable automatic generation of concrete workloads.

7.1.1 Early Validation of FW-based Power Management Strategies

This section introduces the proposed approach for FW-based power management validation using constrained random techniques. At first, the overall workflow is described. Then, we present the specification principles for workload scenarios using constraints. Finally, the developed constrained random generator for these workload scenarios is introduced.

7.1.1.1 Overall Workflow

The overall workflow of the proposed approach is depicted in Fig. 7.1 and detailed in the following. The approach starts with a set of workload scenarios that have been formulated by the user (e.g. system architect or power validation engineer). The scenarios should have different characteristics of power consumption to ensure the thoroughness of the validation. Each scenario is described by workload constraints together with its power and performance budget. The workload constraints define the set of possible legal concrete workloads. Each scenario is furthermore associated with a number N—the minimum number of concrete workloads to be exercised in this scenario. The power and performance budget specification can be either absolute (i.e. absolute power consumption in μW or execution time in μs) or *relative*. Since the former is rather straightforward, in the following, we focus only on the latter (relative) for a more compact representation. Also, in many cases, as the concrete workloads can be strongly varying, it might be not appropriate to specify an absolute budget. The relative budget is specified by percentages of the maximum possible for a concrete workload, e.g. performance within 70% of the maximum but power consumption not more than 50%. This maximum will be calculated by the approach as described below.

Our approach processes each scenario individually. Since the scenarios are independent, it is possible to distribute the computation over a cluster to speed up the overall validation process. In the first step, a *Constrained Random Generator* (CRG) is instantiated. Then, for each scenario, the workload constraints are fed into the CRG, which is then instructed to solve the constraints and generate N different solutions. If the number of solutions is less than N, this is reported back to the user.[1] Each solution of the workload constraints is a concrete application workload for the considered scenario.

Then, for each concrete workload, our approach, in particular the program generator, generates two different programs. These programs are to be executed in a FW/VP co-simulation on the target VP. While they are equivalent from the functional point of view, their power consumptions will be different: While the

[1] This step is omitted from Fig. 7.1 for the simplicity of representation.

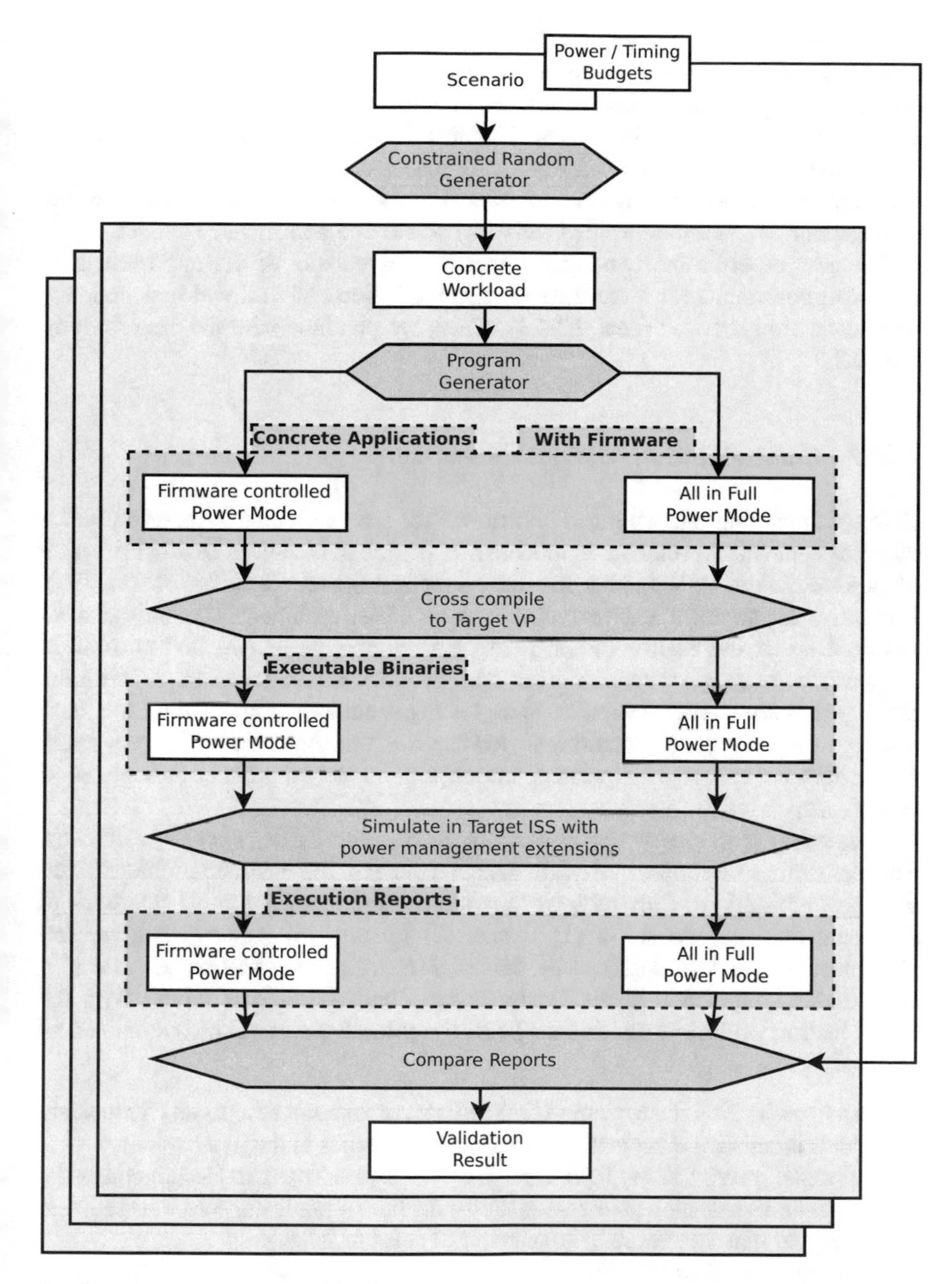

Fig. 7.1 Power management validation overview

first program uses the FW-based power management strategy under validation, in the second program, all components of the VP are set to work in full power mode (i.e. without FW-based power management). The attentive reader will already have deduced that the second program will be used to calculate the maximum power consumption and performance for the concrete workload. Subsequently both programs are cross-compiled to the instruction set of the target VP. The resulting binaries are loaded into the VP and executed. Under the assumption that the VP is *power-aware* and can generate detailed reports on power consumption and performance, these reports are inspected by our approach to validate whether the FW-based power management satisfies the specified (relative) power and performance budget.

We now describe the two most important ingredients of the approach: how workload constraints for a scenario are specified (Sect. 7.1.1.2) and how concrete workloads are generated (Sect. 7.1.1.3). Then, we continue with the case study in Sect. 7.1.2.

7.1.1.2 Constraint-based Workload Scenarios

Before dealing with the workload constraints for a scenario, let us focus on how a concrete application workload is modeled. A workload is viewed as an abstraction of an execution of a SW application and contains a list of *instruction blocks* (IBs). Currently, our approach supports three types of IBs: arithmetic, memory, and IO device. Besides the basic common fields, e.g. block type *IB.type* and position in the list *IB.pos*, every instruction block has specific options, for example, with the block type is arithmetic, the options *num_instr* (i.e. number of instructions in block) and *op_type* (i.e. type of operation) are available. These options describe how many instructions are executed and what operation, e.g. integer addition or multiplication is used, respectively.

A scenario is then a symbolic description of a family of concrete workloads. The constraints describe the relationships between the instruction blocks and their specific options. Currently, we provide the following primitive functions to formulate these relationships: (1) Exists, (2) Ensure, (3) Assert, (4) Size, and (5) Select. Formally a solution, i.e. list of NB instruction blocks, is valid if it satisfies the conjunction of all Exists, Ensure, and Assert constraints. With the list of instruction blocks denoted as LIB, the primitive constraints are defined as follows:

- *Exists(pred)* : $\exists b \in LIB : pred(b)$. The Exists constraint accepts an IB predicate *pred*. It is satisfied if such an IB satisfying *pred* exists in the list.
- *Ensure(sel, pred)* : $\forall b \in LIB : sel(b) \implies pred(b)$. The Ensure constraint accepts an IB selection predicate *sel* and a further IB predicate *pred*. It is satisfied if every IB that satisfies *sel* (i.e. selected IB) also satisfies *pred*.
- *Ensure(sel1, sel2, pred)* : $\forall b_1, b_2 \in LIB$ with $b_1 \neq b_2 : sel1(b_1) \wedge sel2(b_2) \implies pred(b_1, b_2)$. This extended form of Ensure has the same semantics as the simple form, but works for a pair of IBs instead of one single IB.
- *Assert(expr)*. The Assert constraint expects a Boolean expression as argument and is satisfied if the expression is valid.

- *Size(pred)*: $x = |\{b \in LIB | pred(b)\}|$. The *Size* function returns a new symbolic variable x that represents the number of IBs that satisfies the predicate *pred*. The result of the *Size* function can be used to build larger predicates, which can then be passed to either *Exists*, *Ensure*, or *Assert* constraints.
- *Select(pred)*. This helper function provide a way to define named predicates that can be reused in other constraints.

In addition, *Exists* can also be assigned to a named predicate allowing more succinct constraint specification. The predicates are mainly defined using the *lambda* notation, e.g. *lambda* x : $x.type == arithmetic$. is satisfied by any arithmetic instruction block.

Example Constraints An example for a constraint-based workload scenario is shown in Fig. 7.2. The example describes an abstract (symbolic) application workload that starts with CPU-intensive code followed by mixed instructions. The first two lines specify that the initial five instruction blocks have arithmetic type. The next two lines require two of these five arithmetic instruction blocks to be executed with high interrupt frequency from peripherals and IO devices. Line 7 and 8 specify that IO devices are not accessed immediately one after another, there is always processing time in-between. The next two lines require that at least two memory

```
1  // initial five instruction blocks have arithmetic type
2  A = Select(lambda x: x.pos <= 5)
3  Ensure(A, lambda x: x.type == InstrType.Arithmetic)
4
5  // two of the blocks from A have a small irq scaler (frequency at which interrupts
       arrive) - please note numbers starting with '0x' are in hexadecimal format
6  B = Select(A, lambda x: x.irq.scaler != 0 && x.irq.scaler <= 0x50)
7  Assert(Size(B) == 2)
8
9  // the position of DeviceIO blocks is an odd number
10 C = Select(lambda x: x.type == InstrType.DeviceIO)
11 Ensure(C, lambda x: x.pos & 1)
12
13 // at least to memory intensive instruction blocks are available
14 D = Select(lambda x: x.type == InstrType.Memory)
15 Assert(Size(D) >= 2)
16
17 // at least on block has arithmetic type and more than 10000 instructions
18 Exists(lambda x: x.type == InstrType.Arithmetic && x.arithmetic.num_instr > 10000)
19
20 // there shall be an IO-device access with fast processing and another with slow
       processing of incoming/outgoing data (device.scaler denotes the processing
       speed) and the fast block appears before the slow one (specified by last
       constraint)
21 E = Exists(lambda x: x.type == InstrType.DeviceIO && x.device.scaler < 0xfff)
22 F = Exists(lambda x: x.type == InstrType.DeviceIO && x.device.scaler > 0x7ffff)
23 Ensure(E, F, lambda a,b: a.pos < b.pos)
```

Fig. 7.2 A constraint-based workload scenario

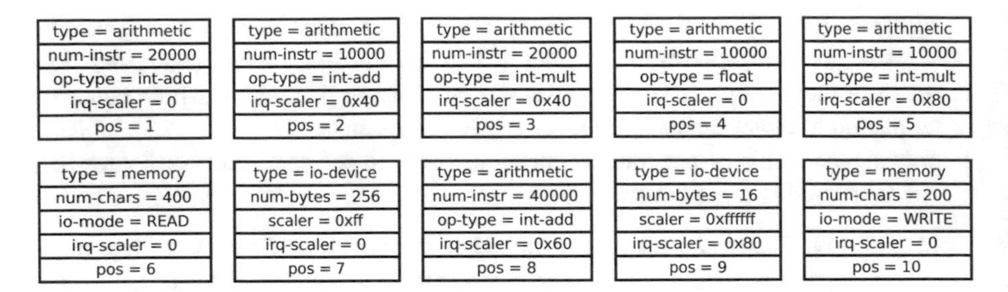

Fig. 7.3 Application workload (constraint solution)

intensive instruction blocks are present in the list. Line 13 requires that at least one arithmetic block exists with more than 10,000 instructions. The last three lines ensure some specific instruction blocks are present. In particular, there shall be an IO-device access with fast processing and another with slow processing of incoming/outgoing data. The last constraint determines the order of these IO accesses: fast before slow.

A solution to these workload constraints with 10 instruction blocks, i.e. a concrete workload of the example scenario is shown in Fig. 7.3. It satisfies all workload constraints, i.e. Assert, Exists, and Ensure constraints. For example, the first five blocks are of arithmetic type. Other options that have not been constrained such as the type of arithmetic operation *op-type* (e.g. integer multiplication or floating point operation) are randomly generated. The generated list of instruction blocks allows the program generator to build a concrete application by randomizing the instructions within a block according to the block properties. The concrete application can then be simulated on the target VP.

7.1.1.3 Constrained Random Generator

The constraint language described above is implemented as a Domain Specific Language in Python (version 3). For a fully integrated SystemC-based flow, it would be better to build the language on top of a CRG framework for SystemC/C++ such as CRAVE [77]. A further advantage is that one could benefit from sophisticated CRG algorithms already implemented in such framework (see, e.g., [132]). However, this would require substantial more implementation efforts. Furthermore, state-of-the-art CRG for SystemC/C++ only supports constraints on bitvectors, while our workload constraints can be formulated more naturally and efficiently on integers (not to be confused with their representation as bitvectors). Therefore, for rapid prototyping and exploring our ideas, we decided to use Python at this stage and leave the option of a SystemC/C++ implementation for the final stage. The CRG as well as the overall flow is implemented completely in Python3. For the generation of concrete workloads of a scenario, the CRG starts with a predetermined number *NB* of symbolic instruction blocks. Then, it initializes *NB* symbolic blocks and maps the specified constraints for these blocks to the SMT fragment QF_LIA (i.e. quantifier-free linear integer arithmetic). In the next step, the CRG employs the state-of-the-art QF_LIA solver Z3 (v4.5.0) to solve the resulting SMT formula and generate N different solutions. Note that Z3 can by default generate only one solution, our CRG contains an all-solution-solving layer over Z3 that adds additional constraints after the generation of a solution to block this solution from being considered in the future. If less than N different solutions can be found, the generator will increase NB to generate more solutions until the number N is reached.

In the following we demonstrate the proposed approach for a LEON3-based VP.

7.1.2 SoCRocket Case Study

This section presents a case study where we apply the proposed approach to the open-source VP SoCRocket [189]. Since there is no freely available VP with power modeling and estimation and SoCRocket is open source, we first extend the base VP with power management features (see Sect. 7.1.2.1) and implement a FW-based dynamic power management strategy (see Sect. 7.1.2.2). Then, in Sect. 7.1.3 we demonstrate our approach by providing an extensive validation of the power management strategy.

7.1.2.1 Power Management Extensions

SoCRocket already includes basic power models. There are three types of power consumption values for each component: (1) static power, (2) internal power, (3) switching power. Both static and internal power can be considered application independent, thus their value only depends on the simulation time. Switching power will increase when the component is actively working, e.g. it depends on the number of executed instructions of the CPU and the number of bytes accessed by the memory and so on. Every component in SoCRocket possesses these power information, see [112] for more information. Adding up those power values for all components allows to compute the total power consumption at every simulation time step. We extend this basic scheme to support multiple power states and discuss how this information can be tracked. Furthermore, we add a lightweight *Power Interface Unit* (PIU) connected to the AHB bus of the system to act as a power interface for the firmware.

Power Modeling For power modeling we add a power layer for every component. This layer stores the power states the component supports together with the currently active state and component specific delays due to power state changes. For example the CPU supports the full power mode (RTM), some power save modes (PS0, PS1, PS2) where it is still able to execute instructions, and sleep modes (DS0, DS1, DS2) where the CPU only waits for interrupts. In general every component supports some power save and sleep states in addition to the obligatory full power mode. For every power state the CPU power layer specifies how many extra cycles are added during instruction processing compared to the base value provided by SoCRocket. The memory power layer specifies how many extra cycles are necessary due to power save modes to process read/write instructions and so on. Similarly the base power consumption values for static, internal, and switching power (which are provided in SoCRocket already) are modified based on the active power state of the components. As example the static power retrieval is shown in Fig. 7.4. Based on the current state s of the component pm a scaling factor is retrieved from a lookup table (Line 2) and applied to the base static power of the component (Line 7).

Power Tracking For power tracking, every component is registered in the power monitor before simulation. Tracking static and internal power is straightforward because it is application independent. Therefore, at registration the static and internal power for every supported power state of every component is retrieved as shown in Fig. 7.4 and dumped to a log file. These power values describe how many static and, respectively, internal power is consumed by the component per second in each power state. Switching power depends on the application code and, therefore, is periodically read and reset for every component, as shown in Fig. 7.5. In order to compute the total power at each time step, the power monitor also dumps all power state changes of every component together with a simulation timestamp.

```
1  virtual double get_sta_power(PM_STATE s) {
2    std::map<PM_STATE, double>::const_iterator it = int_power_coefficients.find(s);
3
4    if (it == int_power_coefficients.end())
5      throw std::runtime_error("Unknown power state (get_sta_power)");
6
7    return it->second * pm->int_power;
8  }
```

Fig. 7.4 Retrieve the static power of a component based on its current power state

```
1   virtual double get_and_reset_swi_power(PM_STATE s) {
2     std::map<PM_STATE, double>::const_iterator it = swi_power_coefficients.find(s);
3
4     if (it == swi_power_coefficients.end())
5       throw std::runtime_error("Unknown power state (get_swi_power)");
6
7     double ans = it->second * pm->swi_power + pending_state_change_power;
8     pm->reset_swi_power();
9     pending_state_change_power = 0;
10    return ans;
11  }
```

Fig. 7.5 Retrieve and reset the switching power of a component based on its current power state

Power Interface Unit The Power Interface Unit (PIU) acts as a power interface for the firmware. Therefore, it provides memory mapped addresses to the firmware, which the firmware can write and read. The PIU is connected ot the AHB bus of the system. Every component is registered in the PIU. The PIU has two tasks: (1) decode firmware commands for power state changes and sent it to the corresponding component and (2) provide hardware performance characteristics to the firmware. In particular the CPU and memory controller track their idle and active times, i.e. their duty cycle. The firmware can access this information through the memory mapped addresses of the PIU.

7.1.2.2 Firmware-Based Power Management

The power layer on top of SoCRocket does not have any logic to decide power state changes of the components. The power management strategy is completely implemented in firmware. Therefore, the firmware manages a set of data structures. Essentially, it is the current power states of the components as well as counters and auxiliary data structures for guiding the power management and synchronizing firmware code called from application code and firmware code asynchronously triggered by interrupts. In the following we describe the duty cycle based power management strategies for the CPU and memory controller as well as firmware code to access IO devices in more detail.

Duty Cycle Based Power Management The power state transitions of the CPU and memory controller are based on duty cycles (i.e. active and idle times) obtained from the hardware. Therefore, the PIU periodically triggers an interrupt. The interrupt handler is shown in Fig. 7.6. It retrieves the duty cycle bitvector from a memory mapped address and updates the power states of the CPU, memory, and IO devices. A duty cycle of 75 for the CPU means the CPU spends 75% of the last time interval being active and was, therefore, idle 25% of the time.

As an example we will describe how these duty cycles are used in the CPU power management strategy in the following. The power management strategy of the memory controller and memories is similar to that of the CPU. For IO devices we use a strategy that will sent them to sleep mode in case they are currently not in use (i.e. there is no pending operation by the application code, which has been interrupted by this interrupt handler) and have not been used in the last time interval.

```
1 void pm_irq_handler(int irq) {
2   uint32_t dc = *DUTY_CYCLE_ADDR;
3   update_leon3_power_state(CPU_DUTY_CYCLE(dc));
4   update_memory_power_state(MEMORY_DUTY_CYCLE(dc));
5   update_devices_power_state();
6 }
```

Fig. 7.6 Regularly triggered by interrupts to update power states of the hardware components

```
1  void update_leon3_power_state(uint8_t leon3_dc) {
2    switch (leon3_stat.pm_state) {
3      case PM_STATE_RTM:
4        if (leon3_dc < 75)
5          leon3_change_power_state(PM_STATE_PS0);
6
7        break;
8
9      case PM_STATE_PS0:
10       if (leon3_dc < 50)
11         leon3_change_power_state(PM_STATE_PS1);
12       else if (leon3_dc >= 75)
13         leon3_change_power_state(PM_STATE_RTM);
14       break;
15
16     case PM_STATE_PS1:
17       if (leon3_dc < 25)
18         leon3_change_power_state(PM_STATE_PS2);
19       else if (leon3_dc >= 50)
20         leon3_change_power_state(PM_STATE_RTM);
21       break;
22
23     case PM_STATE_PS2:
24       if (leon3_dc >= 25)
25         leon3_change_power_state(PM_STATE_RTM);
26       break;
27
28     default:
29       assert (0 && "unkonwn power state");
30   }
31
32   if ((leon3_stat.pm_state == PM_STATE_RTM) && (leon3_stat.num_rtm > 3)) {
33     leon3_change_power_state(PM_STATE_PS1);
34   }
35
36   if (leon3_stat.pm_state == PM_STATE_RTM)
37     ++leon3_stat.num_rtm;
38   else
39     --leon3_stat.num_rtm;
40 }
```

Fig. 7.7 Update CPU power state based on the CPU's duty cycle—regularly triggered by interrupts

Figure 7.7 shows the LEON3 power management strategy (i.e. the CPU core). The strategy will slowly increase the power save modes in case the CPU is idle, but it will immediately go into full power mode when sufficient work is available (*"on-demand"* policy). For example, consider the case in Lines 17–21. The CPU is already in PS1 mode. The action depends on how much time the CPU has been idle in the last time interval:

- at least 75% : the CPU will go into PS2.
- between 75% and 50% : the CPU will stay in PS1.

- less than 50% : the CPU will change into full power mode (RTM).

The Lines 32–39 ensure that the CPU will not stay for too much time in full power mode. In case the CPU does not become idle within three time intervals, it will change to a power save mode to avoid extensive power consumption (and also heat dissipation).

Read IO data The application code does not access IO devices (e.g. peripherals, UART, etc.) directly but only through a function layer provided by the firmware. For example to read 12 bytes from IO device 1, the application code will call io_device_read_data(**int** device_id, **char** ∗dst, **int** num) with device_id=1, num=12, and provide a char pointer *dst* to store the bytes. Figure 7.8 shows the code to read data from an IO device. Interrupts are disabled while accessing data structures shared with the interrupt handlers. The function *leonbare_disable_traps* disables interrupts and *leonbare_enable_traps* re-enables them again.

The function will power up the IO device if necessary, i.e. the function *io_ensure_power_up* will power up the IO device in case it is currently in sleep mode (this happens when the IO device is not used for some time intervals), and then iterate until to_recv chars have been received into *buf*. In each iteration the firmware will try to receive a single char (Lines 10–12) or put the CPU to sleep mode in case no data is available (Line 14).

The *io_wait_for_data* function puts the CPU to sleep mode. Please note, that this is a shared operation with the interrupt handler who also updates the power state of the CPU. Therefore interrupts are disabled before sending the CPU to sleep mode. In SoCRocket we ensure that interrupts are automatically re-enabled when the CPU

```
1  void io_device_read_data(int device_id, char *buf, unsigned int to_recv) {
2    uint32_t n = leonbare_disable_traps();
3    io_device_stats[device_id].pending_io = 1;
4    leonbare_enable_traps(n);
5
6    io_ensure_power_up(device_id);
7
8    unsigned int num_recv = 0;
9    while (num_recv < to_recv) {
10     if (io_get_available_chars(device_id) > 0) {
11       buf[num_recv] = io_read_char(device_id);
12       num_recv++;
13     } else {
14       io_wait_for_data(device_id);
15     }
16   }
17
18   n = leonbare_disable_traps();
19   io_device_stats[device_id].pending_io = 0;
20   leonbare_enable_traps(n);
21 }
```

Fig. 7.8 Firmware function to read data from an IO device

goes into sleep mode. Otherwise the CPU would not wake up again as no interrupts would come in.

7.1.3 Results

In this section we present results of applying our validation approach on the SoCRocket VP. Validation results for 5 scenarios are presented in Table 7.1. For every scenario we generate 50 concrete workloads using our constrained random technique. For every concrete workload a concrete application is generated and executed in full power mode (RTM) and with firmware-based power management on the SoCRocket platform. Table 7.1 shows the average results over all runs. All power consumption values are specified in micro Joule (μJ) and simulation time in seconds. On average 8,000,000 instructions are executed on the SoCRocket platform per concrete workload. Validation of a scenario takes 15 min. in average. All experiments have been run on a Linux machine with a 2, 4 GHz Intel and 16 GB RAM. Please note, with simulation time we do not refer to the wall time, but the time it takes for the application code to execute on the SoCRocket platform, i.e. when the code will run on the real hardware (estimated at system level using the SystemC-based VP). Therefore, a higher simulation time directly implies a lower performance of application code. We define the power and performance budget of the firmware-based power management to be 80% of power consumption and 150% of simulation time compared to full power mode, i.e. the power management should save at least 20% power and should reduce the performance by no more than 50%.

Table 7.1 shows the scenario name in the first column. The second and third columns show results for simulating the concrete application in full power mode (RTM) and with firmware-based power management, respectively. For both modes we further report the total power consumed (which consists of static, internal, and switching power), as well as the simulation time on the SoCRocket platform. The fourth column shows the difference in power consumption and simulation time on the SoCRocket platform between both modes. For example, it can be observed that

Table 7.1 Experiment results (simulation time in seconds, power consumption in μJ) on the SoCRocket platform

	Full power mode		Firmware-based		Difference	
Scenario	Power	Time	Power	Time	Power	Time
(S1) High CPU load	254,969	2.03	94,405	2.42	−62.97%	+19.21%
(S2) Interrupt intensive	161,274	1.09	129,345	1.68	−19.80%	+54.13%
(S3) Alternating workload	397,375	2.77	210,988	3.74	−46.90%	+35.02%
(S4) Memory and IO intensive	1,004,561	7.63	278,397	10.88	−72.29%	+42.60%
(S5) Small tasks	270,656	1.82	208,755	2.88	−22.87%	+58.24%

On average 8,000,000 instructions executed on the SoCRocket platform per concrete workload

the firmware-based power management strategy on average saves 62.97% power consumption at the cost of losing 19.21% performance compared to full power mode for the concrete workloads in scenario S1. In particular, we consider the following scenarios:

S1 describes workload that is very CPU intensive. It generates instructions with high CPU load.
S2 generates interrupt intensive workload. Application code is interrupted by incoming interrupts with very high frequency.
S3 describes workload with alternating instructions blocks. It ensures the neighboring code blocks do not have the same instruction type.
S4 generates workload that is very memory and IO intensive. The CPU load is comparatively low. It ensures that all IO devices are used with different processing speed.
S5 describes workload with many small tasks. This leads to application code with many small blocks of different instruction types.

It can be observed that the power and performance budgets are satisfied for most scenarios (S1,S3,S4). Scenario S2 slightly exceeds both the power and performance budget, and S5 the performance budget only. S2 describes interrupt intensive code which will interrupt the normal application flow and perform some computation before giving the control back. This can lead to inefficient sleep intervals when waiting for IO, as the CPU will wake up from the interrupt (and also check the device for available input again) to be put back to sleep again. S5 changes the workload type very frequently. When changing from CPU intensive to IO or memory bound code and vice versa, the firmware will reduce the power state of the CPU and power it back up again. Therefore, both S2 and S5 can lead to an increased switching frequency of power states in firmware code. We can use our approach to generate additional workload for further investigation. Furthermore, we can plot power diagrams which show the power consumption (of the whole system or any particular component) at different simulation time steps to get further insight.

7.1.4 Discussion and Future Work

Our proposed approach employs constrained random (CR) techniques to generate concrete workloads and then validate that available power and performance budgets are satisfied, by using a power-aware simulation on the VP. First experiments with the LEON3-based SoCRocket VP demonstrate the applicability and effectiveness of our approach. For future work we plan to investigate power-aware coverage metrics in order to evaluate the thoroughness of the validation. A viable solution might be to combine information about reached power states of all components of the system as well as transitions between power states with well-known code coverage metrics. To speed up the overall validation, we plan to distribute the processing of scenarios using multiple threads or even clusters connected by a network. This is a natural and very promising extension as all scenarios are processed independent of each other. Finally, we plan to extend our constraint language and program generator to support generation of multi-threaded workload.

7.2 Maximizing Power State Cross Coverage

This section presents a complementary coverage-driven approach for validation approach for FW-based PM. The main contribution is a feedback-directed workload generation algorithm that generates testcases in an automated manner in order to maximize the cross coverage of power states.

We are not aware of any other coverage-driven validation approach for FW-based PM. A feedback-directed algorithm for maximizing cross coverage of power states is, therefore, to the best of the authors knowledge, completely novel.

In the area of functional verification, automated coverage-driven verification has been one of the most important research topics and thus received much attention. Current state-of-the-art approaches, pioneered by IBM Research [56], are also based on a feedback-directed loop from achieved coverage to the scenario generator. They often integrate machine learning techniques, such as Bayesian networks or data mining, to infer the assumed relationship between coverage and scenario generation directives. We refer the reader to an excellent survey [115] on this topic for more details. These approaches are powerful but generic. The specific context of FW-based PM allows us to develop a specialized, much simpler feedback-directed algorithm. It would be very interesting to compare the performance between the proposed approach and those generic solutions. Unfortunately, they are not publicly available. A more recent and very promising coverage closure approach is based on assertion mining [144]. However, it requires an output-directed notion of coverage that is not directly transferable to our context.

Formal techniques have been shown to be feasible for the *verification* of FW-based PM, see, e.g., [40]. The problem being solved is, however, completely different from validation. Such approach aims to ensure the correctness of a concrete HW/SW implementation of a PM strategy (and thus can only be applied very late in the design flow), while we want to evaluate the effectiveness/soundness of this strategy from the system-level perspective.

We consider systems that use a FW-based PM, that controls the PM strategy (i.e. the active power state of each HW component) in FW based on the observed duty cycles of the HW components (i.e. how long has the component been active and idle, respectively), as has been described in Sect. 7.1.2.2. In the following we start by introducing the concept of a cross power FSM (*Finite State Machine*). Then we present our coverage-guided approach for maximizing power state cross coverage.

Cross Power FSM

A power FSM, e.g. as shown in Fig. 7.9, can be naturally extended to a cross power FSM. Essentially, the cross FSM is a product machine of the N individual power

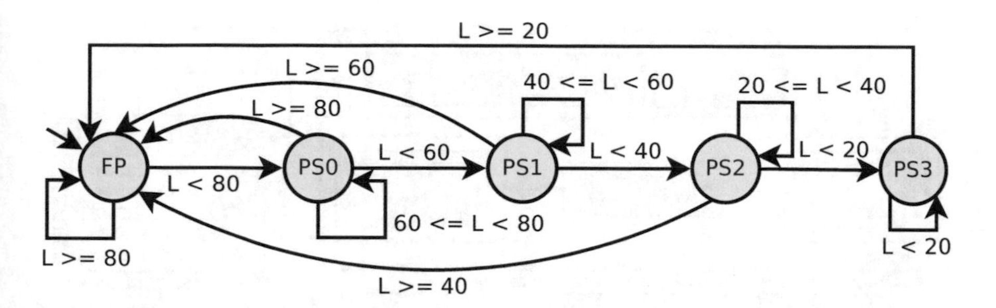

Fig. 7.9 Example power FSM of a component implementing an *on-demand* PM strategy. L denotes the duty cycle (i.e. load) of the component in the last period (time between FW update cycles)

FSMs. A transition between two states $S = (a_1, \ldots, a_N)$ and $G = (b_1, \ldots, b_N)$ in the cross FSM is possible, if the transitions $(a_1, b_1), \ldots, (a_N, b_N)$ are defined in the individual power FSMs, respectively (where $a_1,\ldots,a_n$ and $b_1,\ldots,b_N$ are power states from the corresponding individual FSMs). The duty cycles annotated to a cross transition are simply a combination of the annotations of the individual FSMs transitions. For example, consider a cross FSM, which is a combination of three FSMs, all three as defined in Fig. 7.9. This cross FSM has an edge from state $(PS0, PS1, FP)$ to state $(PS0, PS2, PS0)$ annotated with the load expression $(60 \leq L1 < 80$ and $L2 < 60$ and $L3 < 80)$ where L1, L2, and L3 are the load (i.e. duty cycle) annotations from the three individual FSMs.

7.2.1 Maximizing Power State Cross Coverage

As already mentioned, we assume that the PM strategy is implemented in FW and updated periodically based on the duty cycles of the components in the system.

First we select a number of components from the system and create the cross power FSM based on the individual power FSMs (see Sect. 7.2 for more details on this preliminary step). The individual power FSMs where the (directed) edges are annotated with transition duty cycles (e.g. as shown in Fig. 7.9) can be obtained from the specification or extracted from the FW for each component. Then, we generate testcases that will maximize the coverage of the cross FSM states. We describe our proposed approach in the following.

7.2.1.1 Overview

An overview of our approach is shown in Fig. 7.10. Our approach works in two phases: first a bootstrap phase (upper part of Fig. 7.10) is executed to obtain preliminary coverage information and then a coverage-loop phase to close the

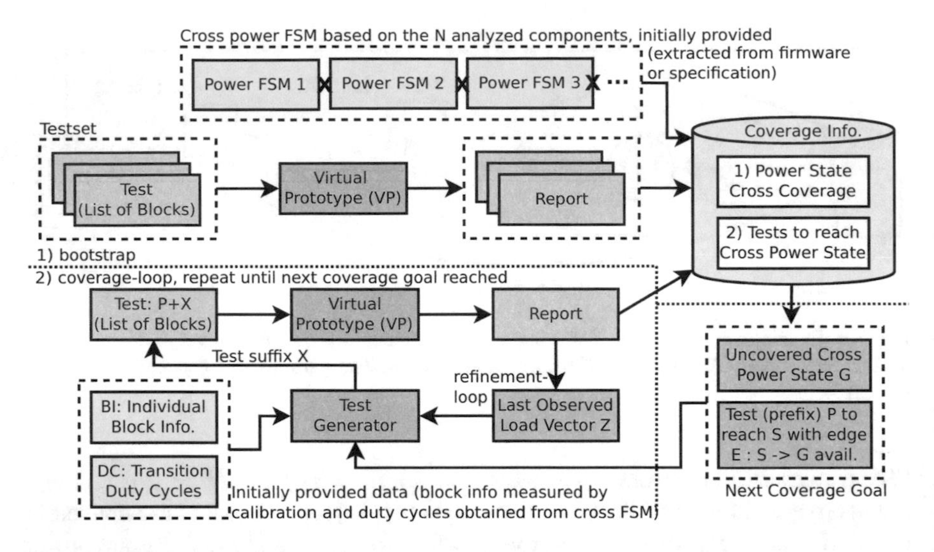

Fig. 7.10 Overview of our approach for test-case generation

remaining coverage gaps (lower part of Fig. 7.10). We keep track of relevant coverage information in a shared data structure (right side of Fig. 7.10). In the following we describe our approach in more detail.

In the *bootstrap phase* a set of testcases one after another is linked with the FW, then cross-compiled and executed on a power-aware VP. These testcases can be obtained in various ways, for example, created by a testing engineer or using (constrained) random generation techniques. We assume that each testcase consists of a list of blocks (i.e. the SW workload). A block is a list of instructions, potentially including (bounded) loops. Each block corresponds to a specific instruction execution profile (e.g. memory, arithmetic, sleep, etc.) and runs only for a short amount of time compared to the FW-based PM update cycle interval. The number of blocks can be randomly chosen, but it should be long enough to trigger (preferably multiple) PM update cycles to obtain useful coverage information. For each execution an (execution) report is obtained. The report contains various informations about the test-case execution, including the execution time of each block and the power state changes of every component in the system during the PM update cycles.

A shared data structure stores the relevant coverage information from the reports between all execution runs. Essentially, these are two pieces of information: (1) Based on the observed power state changes in the report, the visited cross power states are marked. This information is used to select the next uncovered goal state G. (2) A mapping W from cross power state S to list of blocks B (i.e. testcase or prefix of a testcase) is updated, such that the execution of B will lead to the cross power state S. This information is used to generate a prefix of blocks to reach a specific power state.

The *coverage-loop phase* starts after all testcases of the bootstrap phase have been executed to close the remaining coverage gap. It will consider all edges `E = S -> G` from the cross FSM, where the start S is covered and the goal G is not, one after another. Based on the mapping W from the shared coverage data, a prefix P of blocks is available whose execution will reach the state S. Thus, it is only necessary to generate a suffix X in order to hit the (cross) power transition `S -> G`. Therefore, the test generator first extracts the goal load from the cross FSM. The goal load is an interval vector (GLIV), where each interval denotes the expected duty cycle for every component in the cross FSM to apply the cross power transition `S -> G`. For example, consider a cross FSM for three components, each using Fig. 7.9 power FSM individually, with the start state S=(PS0, PS1, FP) and the goal state G=(PS0, PS2, PS0). Then the goal load (cuboid) is defined as ($60 \leq L1 < 80$ and $L2 < 60$ and $L3 < 80$) where L1, L2, and L3 are the duty cycles (i.e. loads) of the three components. Based on the GLIV (i.e. goal load), the suffix X of blocks (long enough to reach at least one PM update cycle in order to have an effect) is generated. As already mentioned, this suffix X should consist of blocks such that execution of (P + X) will trigger the cross power transition from S to G, i.e. reach the expected goal load in the next PM update cycle. Figure 7.11 shows the principle.[2] After executing the testcase (P + X) there are two possibilities:

1. The cross state G has been covered. In this case simply proceed to the next uncovered cross state.
2. Otherwise, the suffix X needs to be refined. Refinement is based on the information that has been collected during execution of (P + X). In particular, refinement is based on the load vector Z, which contains the observed duty cycle for each component (obtained from the first update cycle when executing X, which is U4 in Fig. 7.11). Refinement and re-execution will iterate until G is covered or the test generator gives up on refinement, e.g. because the maximum number of refinement steps is reached. In case refinement is not possible, the goal state G might still be covered later, because in the cross power FSM there can be multiple edges with different start states S that reach G.

We provide more information on the test generation process in the following.

7.2.1.2 Coverage-Loop

The coverage-loop test generator starts with a goal load interval vector (GLIV), as described in the previous overview section, and generates a suffix of blocks X. The test generation is based on mixing blocks to obtain specific load intervals for the

[2]The execution of the last block of P can still reach into the execution period of X (i.e. the beginning of the interval between U3 and U4 in Fig. 7.11). However, by keeping the blocks short compared to the length of the periods between update cycles, this last block of P has only negligible influence on the overall execution of X.

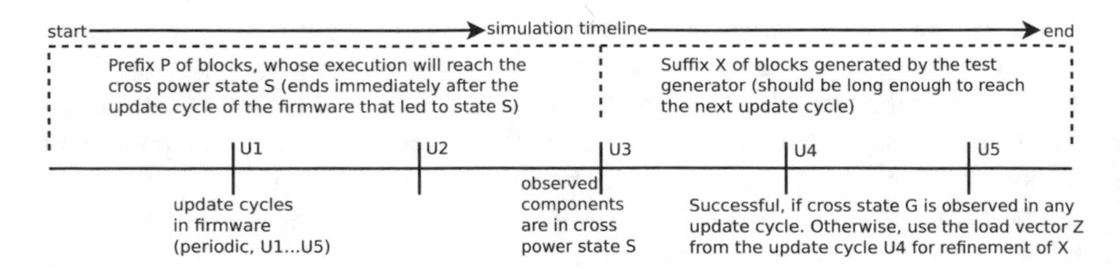

Fig. 7.11 Conceptual overview of the test generation and execution process in the coverage-loop

components within the system. Therefore, a block calibration is performed once before the coverage-loop starts to obtain *individual block information* for every block. Therefore, every (instruction) block is executed individually on the virtual prototype (VP) for a fixed number of times (long enough to trigger an update cycle in FW) while keeping the system in full power mode. By doing so a concrete load vector (i.e. duty cycles of each component of the cross FSM) and the (average) execution time of one individual block is obtained and stored. We perform the calibration only in full power mode to avoid the state space explosion of having to consider the exponential many (in the number of components in the system) different power state configurations of the system. Please note, this calibration has only to be done once. In Sect. 7.2.2 we present concrete blocks and calibration results for our case study.

Based on these individual block information obtained from calibration, the test generator works iteratively through multiple different modes, starting from simple to more elaborate ones. In this work we consider the *point-mode* and *line-mode* generation.

First the *point-mode* generation is applied. Therefore, every individual block load vector V (obtained through initial calibration) is checked against the goal load interval vector (GLIV). If V is contained in GLIV (for example, with three components V would be a point and GLIV a cuboid), the corresponding block B of V matches the goal load. In this case the test generator will generate a suffix X consisting only of B blocks. The mixing vector is defined as [(1.0, B)], i.e. only a single block is used with full weight (i.e. a factor of 1.0). The expectation is that the observed load vector after execution of X will match V and hence cover the GLIV. In case the execution does not match, the next individual block is considered. No refinement is performed in point-mode, since only a single block is used. In case the GLIV cannot be matched using *point-mode*, the generator proceeds to *line-mode*.

In *line-mode* the generator will consider all lines obtained by combination of two block load vectors. For every such (finite) line denoted by two (end-)points L=(V1, V2) the intersection with the GLIV is tested. If there is an intersection, the closest point P on the line L to the center point of the GLIV is computed. The mixing vector M is computed based on the (inverse) distance of P to the edge points V1 and V2 of L. Let d1 and d2 be the distances of P to V1 and V2 with B1 and B2 being the blocks associated with V1 and V2, respectively. Then M is defined as M=[(d2/(d1 + d2), B1), ((d1/(d1 + d2)), B2)]. The division ensures that the factors sum up to 1.0, i.e.

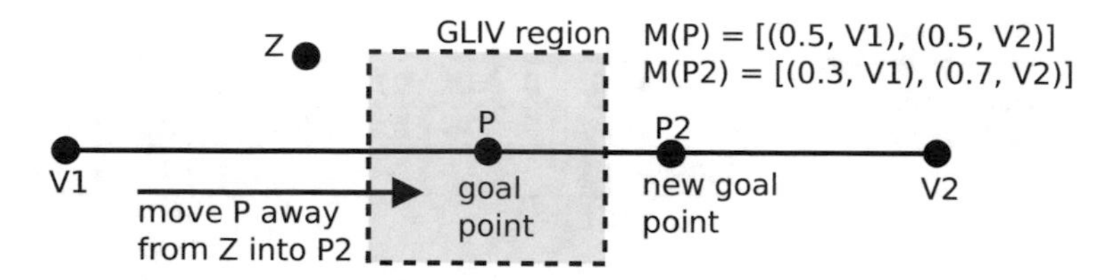

Fig. 7.12 Abstract example demonstrating the line-mode refinement approach in a two-dimensional space

are normalized. The expectation is that the observed load vector Z after execution of X will match the point P and hence cover the GLIV. In case the execution does not match, a refinement approach is started. The refinement approach will modify the block weights of the mixing vector M. Essentially, it will move P in the direction of either V1 or V2, away from the observed vector Z. Figure 7.12 shows the principle. The new goal vector P2 has a stronger influence of V2 (hence its weight factor has increased from 0.5 to 0.7) and thus we expect that the next observed load vector will move towards the original goal vector P and thus be enclosed by the GLIV.

Further modes can be defined, for example, using planes or tetrahedra in combination with barycentric coordinates to allow for a more flexible mixing of blocks. However, our experimental evaluation indicates that using the line-mode can already be sufficient to obtain (close to) maximal power state cross coverage. Therefore, we leave it for future work to implement and evaluate more sophisticated mixing modes and refinement procedures. In the following we describe how to transform a mixing vector to a concrete list of blocks (i.e. the suffix X).

7.2.1.3 Final Test Generation

Given a mixing vector of blocks, for example, M=[(0.7, A), (0.2, B), (0.1, C)], the final task is the generation of the actual suffix X of blocks. This happens in two steps:

First the factors are re-scaled by the average individual block execution time, which has been obtained through the initial calibration. The reason is that some blocks might execute for a (significantly) longer time than others, thus it is necessary to divide the factors by the execution time to keep the proportion of the blocks in the final suffix X intact. After re-scaling, the factors are normalized again to sum up to 1.0 and passed to the next step.

Before explaining the next step, please note that the block execution time can deviate at runtime compared to the calibration result. The reason is that calibration happens in full power mode only, but during execution a PM strategy is active, and caching effects can have an impact on block execution when blocks are mixed. Further, the overall block length need to be long enough to ensure that an PM update cycle is triggered in the FW (otherwise no power state transition happens).

Table 7.2 Example that demonstrates the generation of blocks from a mixing vector by interlacing them

Step		0	1	2	3	4	5	6	7
Budgets	B	0	0.5	0.3	0.1	-0.1	0.4	0.2	0
	C	0	0.6	0.5	0.4	0.3	0.2	0.1	0
Blocks			A	A	A	AB	A	A	ABC

Therefore, the number of blocks generated as suffix should be conservatively (over-) approximated.

However, in doing so, one has to be careful when generating blocks. For example, consider the mixing vector M=[(0.7, A), (0.2, B), (0.1, C)] and assume that 500 blocks should be generated. With an update cycle interval of, for example, 50 ms and a minimal block execution time of 0.2 ms in full power mode this is a valid approximation. But simply generating 350 (0.7*500) A, 100 (0.3*500) B, and 50 (0.1*500) C blocks one after another will not work as expected. The reason is that the execution time of 350 A may already exceed the 50 ms update interval, rendering the mixing invalid (because the blocks B and C will not influence the duty cycles in this update interval).

Therefore, we employ an algorithm to interlace all blocks from the beginning. Table 7.2 shows the result of the algorithm for the first seven steps. The first row lists the step number (zero is the initialization). The second and third rows show the budgets for the blocks B and C. The fourth row shows which blocks are generated in the corresponding step. The algorithm works as follows: First it obtains the block with the highest factor, in this case A with 0.7. All other blocks are associated with budgets initialized to zero. In every step a block of A is generated (since it has the highest factor) and its factor is added to the budgets of all other blocks. Then the budgets of B and C are decremented by its factors, respectively. If the budgets falls equal or below zero, then a block of B or C, respectively, is generated. In this example, after seven steps, the budgets of B and C start repeating, thus the pattern *AAAABAAABC* of blocks is repeated until 500 blocks have been generated.

7.2.2 Case Study

We have implemented our proposed approach in Python. As a case study, we consider the SoCRocket VP [189]. SoCRocket is a power-aware open-source VP written (primarily) in C++ (around 50k lines of code). We have extended SoCRocket to include a lightweight power layer that associates each component in the VP with a power FSM. Further we have added a *Power Interface Unit* (PIU) as described in the preliminary section which allows the FW to control the power transitions of each component in the VP.

The SoCRocket VP consists of various components including a LEON3-based CPU, an interrupt controller, a UART interface, and a memory with corresponding controller, connected by an AMBA-based bus system. For this case study we have added a special processing unit (SPU) which allows to perform special operations independent of the CPU. The SPU is configured through memory mapped (MM) writes. The CPU can either actively wait for the result (spin) of the SPU or sleep until the SPU triggers an interrupt. Further, the SPU is a bus master by itself and thus can independently access the memory.

We consider a three-dimensional cross power FSM combining the FSMs of the CPU, memory (including memory controller), and the SPU, respectively. Thus, our goal load interval vector (GLIV) is a cuboid, representing the duty cycles of these three components. Every component can be either in full power mode (FP) or one of four power safe modes: PS0, PS1, PS2, or PS3, with PS0 being the least and PS3 the most power saving mode. Thus, the complete cross power state space in this case study consists of 5*5*5 = 125 (interdependent) states.

In the following we first introduce the individual (instruction) blocks that we use and provide the pre-computed calibration information. Then we present the results of our experiments using four different PM strategies.

7.2.2.1 Block Definition and Calibration

Table 7.3 shows the individual block informations. The first row shows the load vector measured on the VP for the CPU, memory, and SPU, when executing the block exclusively with all components of the VP being in full power mode (so the FW-based PM strategy is switched of for this measurement). The reason that we do calibration in full power mode only, is to avoid the state explosion of considering all combinations of power states for all different components in the system. The second row shows the average runtime in nanoseconds (NS) for executing only one block.

Table 7.3 Individual block information obtained by running the corresponding block exclusively on the VP (all components in full power mode, i.e. no PM enabled in FW) and measuring the duty cycles (load) and average runtime

	Arithmetic	Arithmetic 2	Memory	Memory 2	Memory 3	Memory 4
Load (CPU, MEM, SPU)	(100, 1, 0)	(20, 81, 0)	(65, 44, 0)	(76, 34, 0)	(51, 54, 0)	(53, 53, 0)
Avg. runtime in NS	465,330	424,360	291,080	384,950	133,850	336,990

	Sleep	CpuSleep NoMem	CpuSleep WithMem	CpuSpin NoMem	CpuSpin WithMem
Load (CPU, MEM, SPU)	(2, 1, 0)	(7, 8, 96)	(3, 86, 96)	(68, 8, 100)	(33, 83, 100)
Avg. runtime in NS	1,069,000	1,123,760	1,104,760	1,083,230	1,155,440

```
1  #define REPEAT_10(x) do { x; x; x; x;
       x; x; x; x; x; x; } while (0);
2  #define REPEAT_100(x)
       REPEAT_10(REPEAT_10(x););
3
4
5  int ArithmeticBlock(volatile int x) {
6    int sum = 0;
7    int i;
8    for (i=1; i<2500; ++i) {
9      sum += i; sum -= x;
10   }
11   return sum;
12 }
13
14 void MemoryBlock() {
15   volatile int a[1024];
16   int k;
17   int x;
18
19   for (k=0; k<10; ++k) {
20     REPEAT_100(
21       x = a[k];
22       a[512+k] = x;
23       a[k] = a[512+k];
24     );
25   }
26 }
27
28
29 void CpuSpinWithMemBlock() {
30   volatile char in[1024];
31   volatile char out[1024];
32
33   *SPU_inputaddr = (uint)&in[0];
34   *SPU_outputaddr = (uint)&out[0];
35   *SPU_operation = 256;
36   *SPU_running = 1;
37
38   // keep spinning until result is
        there
39   while (*SPU_running) {
40     ;
41   }
42 }
```

Fig. 7.13 Example instruction blocks with different type for illustration: an arithmetic, memory, and SPU access block

In total we have defined eleven different blocks. Two blocks that perform arithmetic operations based on addition, subtraction inside of loops. Four blocks that primarily perform various memory operations, including swapping and copying memory elements. A sleep block that will power down the system, hence reducing the duty cycles of the components. Four blocks to interact with the SPU: The CPU can sleep or spin and the SPU can either access the memory to perform the computation or not. When sleeping, the CPU will wake up by an interrupt triggered from the SPU, and when spinning, the CPU will actively keep polling the SPU using memory mapped IO.

For illustration, Fig. 7.13 shows (1) an arithmetic block, (2) a memory block, and (3) a block accessing the SPU. The first block performs addition and subtraction inside a loop. A *volatile int* argument is passed in from the main function (which is just a local variable initialized with a constant value) to avoid pre-computing the result by the compiler due to optimizations. The return value from the function is ignored from within the main function, and it ensures that the computation of the result value is not discarded. In general one has to be careful when writing C code due to compiler optimizations/re-structuring, which can make it more difficult to define suitable blocks. This could be circumvented by writing the block code in assembler directly. Also please note, that writing the blocks has to be only done once. The second block performs multiple memory swap operations. The *REPEAT* macro is used to eliminate some loop checking and update operations, hence putting more weight on the memory operation. Again, to avoid compiler optimizations, the array, that is operated on, is declared volatile. The third block employs the SPU to perform some special operation. The SPU is configured per memory mapped (MM) access to read from and write to a specific memory region. A MM write to the *running* register will start the SPU operation. The CPU is actively waiting (spins) until the result is available (indicated by a zero in the *running* register of the SPU). The SPU is itself a bus master and will perform various memory operations to perform its computation.

In general the blocks should be defined in such a way, to have different load values which cover the cross load state space as thoroughly as possible. This allows to cover the remaining gaps in the cross power state space by mixing the blocks in different combinations. In this work we consider lines between blocks (i.e. their load vectors) for mixing, though this can be further extended to planes or tetrahedra, etc., if necessary. With 11 different blocks we have a total of $\binom{11}{2} = 55$ line combinations. A line describes the load of the three components that can be achieved (in our model, which is a prediction of the actual VP loads) by mixing its endpoints.

7.2.2.2 Experiments

We have evaluated our approach on four different duty cycle based PM strategies. The strategies are implemented in FW and executed periodically during an update cycle:

1. *On-demand*: This strategy will gradually power down the component, but immediately transition to the full power mode when work is available. Figure 7.9 shows the corresponding power FSM.
2. *Conservative*: Will gradually power down (as the *on-demand* strategy) and also gradually power up the component, visiting each power state one after another. Thus, it takes multiple update cycles to fully power up a component which has been in a (deep) power saving mode.
3. *Balanced*: Gradually powers down from FP to PS1 and gradually powers up from PS3 to PS1. From the PS1 state immediately switches to FP or PS3 when work is available or the component is idling, respectively.
4. *Combined*: Use a different strategy for every of the three considered components: *on-demand*, *conservative*, and *balanced* for the memory, SPU, and CPU component, respectively.

In the *on-demand*, *conservative*, and *balanced* setting all three components use the same strategy. We run the experiments on a Linux machine with a 2,4 GHz Intel processor and 32 GB Ram. Table 7.4 shows the results. The right half shows results for the four above-mentioned PM strategies. The table is separated by double lines into two parts and a header.

The upper part (not the header) shows the information of using a random test generator (*Random-only*). The number of blocks is constrained to be between 1000 and 2000 for each test. The blocks itself are currently randomly generated. This table part shows the execution time of all tests together in seconds, the power state

Table 7.4 Experimental results for our approach

Technique to maximize cross coverage			Duty cycle based PM strategy implemented in FW			
			On-demand	Conservative	Balanced	Combined
Random-only		Time in sec.	14,144.81	13,795.99	13,172.07	13,104.20
		coverage	67/125 (53.6%)	27/125 (21.6%)	76/125 (60.8%)	61/125 (48.8%)
		Num. tests	1000	1000	1000	1000
Our approach	(1) Random (bootstrap)	Time in sec.	1450.97	1381.74	1364.74	1338.39
		coverage	55/125 (44.0%)	23/125 (18.4%)	53/125 (42.4%)	49/125 (39.2%)
		Num. tests	100	100	100	100
	(2) Point-mode	Time in sec.	1617.09	1790.34	1680.92	2210.17
		coverage	118/125 (94.0%)	111/125 (88.8%)	113/125 (90.4%)	123/125 (98.4%)
	(3) Line-mode	Time in sec.	3173.76	2698.69	2634.66	1401.87
		coverage	124/125 (99.2%)	115/125 (92.0%)	121/125 (96.8%)	125/125 (100%)
		Total time in sec.	6241.82	5870.78	5680.32	4950.43

cross coverage achieved by executing the tests, and the number of tests generated and executed.

The lower part shows results for our proposed approach. It performs three steps one after another. First it starts by bootstrapping the coverage with random testing. This is the same as the *Random-only* approach but we only use 100 testcases for this bootstrapping. The reason is that adding additional random tests does not increase the coverage very much (as the *Random-only* row has demonstrated). Then our coverage-loop is executed working in *point-mode* and *line-mode* as has been explained in Sect. 7.2.1.2. The test generator is using a suffix X of 600 blocks. For every step we report the execution time in seconds as well as the total coverage obtained after the step. The last row in this lower part of the table shows the total execution time of all steps of our approach together.

It can be observed that random testing does not perform well on this problem instance. For example, increasing the number of tests from 100 (see the bootstrap phase of our approach) to 1000 does increase the achieved coverage only marginally (e.g. from 44% to 53.6% and from 18.4% to 21.6%), even though a large part of the cross power state space is still uncovered, and at the same time the runtime grows (roughly) linearly by a factor of 10. This result demonstrates that an approach performing random test generation is not suitable for our use case.

In contrast it can be observed from the results that our coverage-driven approach works very well to close the remaining coverage gap. Close to 100% power state cross coverage is achieved for every considered PM strategy with reasonable runtime overhead. Already the point-mode strategy is sufficient to achieve very high coverage of around 92% on these examples. Applying the line-mode strategy afterwards does increase the coverage further to up to 100% with an average coverage of 97%. For some PM strategies the coverage is still (slightly) below 100%. In general the reason is that either: (1) a *stronger* mixing model, or (2) different blocks that cover the cross load state space more thoroughly are required. In this work, the reason for uncovered cross states has been the second case, in particular, that we were unable to define blocks with a very high CPU and memory load at the same time (due to synchronization between them). Careful analysis reveals that most of these uncovered cross states are in fact unreachable: 9 out of the 10 uncovered cross states for the *conservative* PM strategy and all 4 uncovered cross states for the *balanced* PM strategy are unreachable. Only one reachable cross state has been missed for the *on-demand* and also only one for the *conservative* PM strategy by our approach. Thus, our approach achieves optimal and near optimal results for the considered PM strategies by employing our mixing model and block definitions. This experimentally proves our approach to be very effective in maximizing power state cross coverage.

7.2.3 Discussion and Future Work

Cross coverage of power states is an effective but at the same time challenging metric to evaluate the quality of a testset in validating power and timing constraints. Scalability can be an issue, because the cross FSM can grow exponentially. However, in general the number of power states per component (and thus single FSM) is rather small. Furthermore, using a subset of (important) components, possibly in different combinations of small cross FSMs, can already provide very useful coverage results. Our experimental evaluation demonstrated the applicability and efficacy of our coverage-driven approach in maximizing the power state cross coverage. Nonetheless, there are still possible directions for further improvements:

1. Currently individual block information (calibration) are obtained with the system running in full power mode to avoid the state explosion of calibrating the system with all possible power state combinations. However, the real system will switch power states during execution and thus deviations from the calibration result can be observed. A viable solution might be to calibrate the system only for the (cross) states of the considered cross FSM. The cross FSM is typically based on a small number of components (three in our case study) and has arguably the biggest influence on the runtime prediction (since the goal is to maximize coverage of the cross FSM).
2. Caching and potentially other side effects, due to block mixing, can lead to deviations of the (static) calibration result at runtime. Thus, it seems useful to integrate dynamic information, observed at runtime during testcase execution, into the test generation process.
3. Integrate more sophisticated block mixing and refinement procedures. Our *line-mode* approach can be extended to, e.g. *plane-mode* or *tetrahedra-mode*, in combination with barycentric coordinates, to allow for a more flexible mixing of blocks. This becomes particularly useful for larger and higher-dimensional cross FSMs.
4. Investigate the use of formal verification techniques at the abstraction level of VPs, e.g. [88, 100], to automatically identify unreachable cross states and to cover cross states that proved to be very difficult to reach with simulation-based methods.

7.3 Summary

In this chapter we presented two complementary approaches for early power validation of firmware-based power management at system level using SystemC-based VPs.

The first approach employs constrained random techniques to generate concrete workloads and then validate that available power and performance budgets are satisfied, by using a power-aware simulation on the VP. The main novelty of the approach is the modeling of workloads based on constrained random techniques. Each workload corresponds to a system-level use case with a specific power consumption profile and is described by a set of constraints. This enables an automated and comprehensive generation of different workloads, hence reducing the risk of missing a corner case scenario.

The second approach works in a coverage-driven fashion with the goal of maximizing the cross coverage of power states. Cross coverage of power states is an effective but at the same time challenging metric to evaluate the quality of a testset in validating power and timing constraints. The coverage metric enables to objectively measure the quality aspect as well as to guide the generation of new workloads. The main contribution of the proposed approach is a feedback-directed workload generation algorithm that generates testcases in an automated manner. In particular, we integrate a coverage-loop that successively refines the generation process based on previous results.

The experiments using the SoCRocket VP demonstrated the applicability and efficiency of our proposed approaches.

Chapter 8
Register-Transfer Level Correspondence Analysis

A VP-based design flow enables HW/SW Co-Design. The VP is used for early SW development and serves as executable reference model for creating and verifying the RTL description of the HW. As already mentioned before, the VP is pre-dominantly created in SystemC TLM. A correspondence analysis between TLM (VP) and RTL (HW) can improve the VP-based design flow by further automating refinement steps and obtain more accurate results on VP-based SW analysis approaches. This chapter presents two novel approaches that perform a correspondence analysis between RTL and TLM to enable the aforementioned benefits.

Section 8.1 presents the first proposed approach that enables an automated TLM-to-RTL property refinement. It enables to transform high-level TLM properties that are used for verification of the VP into RTL properties to serve as starting point for RTL property checking. This avoids the manual transformation of properties from TLM to RTL which is error prone and time consuming. The approach is based on a static transactor analysis based on symbolic execution. The approach has been published in [94].

The second proposed approach performs an RTL-to-TLM fault correspondence analysis to improve the accuracy of a VP-based error effect simulation. An error effect simulation essentially works by injecting errors into the VP during SW execution to check the robustness of the SW against different HW faults. This analysis is very important due to increased risk of faults, e.g. due to radiation, aging, etc., which can lead to an observable error and failure of the system. Error effect simulation with VPs is much faster than with RTL designs due to less modeling details at TLM. However, for the same reason, the simulation results with VP might be significantly less accurate compared to RTL. To improve the quality of a TLM error effect simulation, a fault correspondence analysis between both abstraction levels is required. Section 8.2 presents a case study on applying fault localization methods based on symbolic simulation to identify corresponding TLM errors for transient bit flips at RTL. This approach has been published in [89, 93].

V. Herdt et al., *Enhanced Virtual Prototyping*,
https://doi.org/10.1007/978-3-030-54828-5_8

8.1 Towards Fully Automated TLM-to-RTL Property Refinement

In the VP-based design flow, SystemC TLM models representing the VP are available very early and get successively refined to RTL. TLM models also serve as *reference models* to verify the corresponding RTL models later. The verification is commonly performed by a TLM/RTL co-simulation, where the same input stimuli are applied to both models and their outputs are checked for equivalence. This step requires an additional executable verification component, known as *transactor*, to bridge the function calls at TLM with the signal-based interfaces at RTL and vice versa, as shown in the upper half of Fig. 8.1.

Obviously, the correctness of TLM models is also of great importance. In the past few years, a wide body of verification techniques at TLM has been developed ranging from simulation-based (e.g. [55, 161, 204]) to formal verification (e.g. [38, 72, 82, 86, 88, 90, 134, 135]). Please note that we do not try to be comprehensive but only mention a few representative approaches. Most of these approaches are based on the principle of *Assertion-Based Verification* (ABV) [59], which has been very successful for RTL verification. Assertions, also known as properties, provide a mean to formally capture the functional specification of the design. They specify conditions on design inputs, outputs, and internal variables that are supposed to hold at all times. These conditions can also be temporal, i.e. referring to a sequence of values over time. Assertions can be automatically translated into executable monitors to be co-simulated with the design to check the specified conditions. For the specification of TLM properties, extensions to standardized language such as IEEE-1850 PSL or IEEE-1800 SVA have been proposed [55, 205].

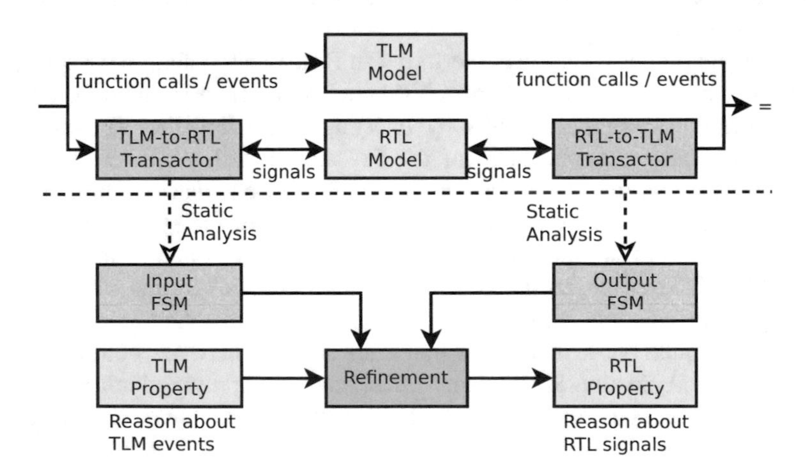

Fig. 8.1 TLM-to-RTL property refinement overview

While the verification step based on TLM/RTL co-simulation is indispensable, the obtained results are far from complete and cannot ensure the absence of bugs at RTL. For some special cases when the TLM models are high-level synthesizable (e.g. when all used constructs are in the *SystemC Synthesizable Subset*), formal sequential *High-Level Equivalence Checking* (HLEC) is supported by existing EDA tools such as Calypto SLEC or Synopsys Hector. A complementary approach to HLEC is provided by leveraging *Path Predicate Abstraction* (PPA) [209]. PPA ensures that the high-level model is a sound abstraction of the RTL implementation and hence enables to establish a well-defined relation between both abstraction levels that can be utilized for verification purposes. A subset of SystemC called SystemC-PPA is provided for the purpose of describing such high-level models with PPA [146].

For general TLM models, it is highly desirable that co-simulation is complemented with other (preferably formal) approaches. A promising direction is to reuse properties that have been proven on the TLM model. Due to the semantic differences of the involved abstraction levels, straight-forward reuse is not possible. The translation process, i.e. TLM-to-RTL property refinement, is mostly manual, therefore error prone and time consuming.

Several improvements to the manual TLM-to-RTL property refinement have been proposed. Ecker et al. [49] formulated a set of requirements for the refinement process. In a follow-up work [199], an automated refinement framework has been introduced. Still, a set of refinement rules have to be defined by verification engineers before the automated translation can start. Chen and Mishra [32] proposed a similar approach that requires a formal (temporal) semantic mapping between TLM functions and clocked RTL signals. Pierre and Amor [170] developed a set of pre-defined "pattern-matching" rules to ensure that the refined RTL properties belong to the *simple subset* of PSL, which is generally easier to verify. However, the need for a manual semantic mapping is still not circumvented. Bombieri et al. [21] proposed to reuse TLM properties in a TLM/RTL co-simulation. Thanks to the availability of transactors, RTL signals over time are converted back to TLM transactions and the TLM properties can thus be checked on the RTL design. This is, however, not a semantic approach, i.e. no corresponding RTL properties can be derived to apply, for example, RTL property checking.

We propose a TLM-to-RTL property refinement approach based on transactors that can operate in a *fully automated* way. The lower half of Fig. 8.1 shows an overview. The main idea lies in a better reuse of the readily available transactors. At the core of our approach is a static transactor analysis based on symbolic execution (Sect. 8.1.2). Essentially, the analysis reverse-engineers the executable transactors to create a formal specification of the underlying protocol as *Finite State Machines* (FSM). Then, TLM properties are refined by relating high-level TLM events with RTL signal combinations at different clock cycles based on the FSM (Sect. 8.1.3). We believe that the proposed approach is of great practical interest. While it is possible to generate transactors automatically from a formal specification [13], in practice, verification engineers still develop transactors manually from scratch based on a textual specification. As the proposed approach can be best explained on

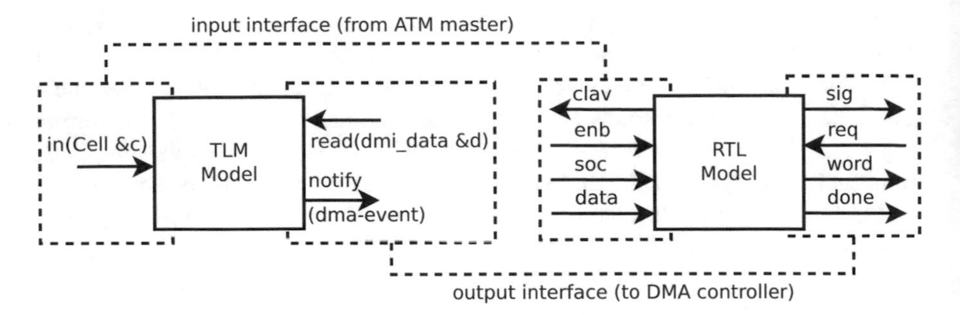

Fig. 8.2 UTOPIA controller TLM and RTL model IO interface (simplified)

concrete examples, we introduce in Sect. 8.1.1 a realistic case study for illustration and also feasibility demonstration.

8.1.1 UTOPIA Case Study

As case study we consider an UTOPIA-based (*universal test and operations PHY interface for ATM)* controller acting as receiver slave device. It communicates with an *Asynchronous Transfer Mode* (ATM) master device using ATM cells, provides an internal buffer to store up to two cells, and a DMA interface to access the buffer. For illustrative purposes, we use a simplified data model, where a cell consists of 3 bytes and the DMA controller word size is 16 bits. DMA access is expected to be word-aligned thus the internal buffer cells are zero-padded (one byte here). Figure 8.2 shows the TLM and RTL IO interface:

TLM Interface The *in* function accepts a whole cell and writes it into the internal buffer. It blocks in case the buffer is full. The (dma-) read function provides a direct memory interface for reading a buffer cell (through a char pointer in *dmi_data*). The (dma-) event signalizes that a cell is available for reading (condition checked after every *in* or *read* call). For this interface, we have five TLM events for property specification: *in:begin*, *in:end*, *read:begin*, *read:end* and (dma-) *event:notified* (abbreviated *notified* as it is the only event).

RTL Interface On the input side *clav* (cell available) signals the controller is able to receive a whole data cell, i.e. buffer not full. The master then starts the transfer by activating the *enb* (enable) signal, sets *soc* (start of cell) to 1 and data to the first byte of the cell. In subsequent clock cycles, the remaining cell bytes are transferred, one after another through *data* signal. During transfer, soc stays 0 and enb stays 1.

On the output side, *sig* (read signal) is active iff a whole cell is available in the internal buffer. DMA controller enables *req* (request) when ready to receive. Then, the data is provided sequentially through the *word* signal (2 byte steps). Finally, *done* is enabled one clock cycle after data transfer has finished.

```
1  struct TLMToRTLTransactor {
2    bool data_available = false;
3    char data[CELL_SIZE];
4    event empty;
5
6    void tlm_input(const Cell &c) {
7      if (data_available) {
8        wait(empty);
9        assert (!data_available);
10     } else {
11       data_available = true;
12       for (i=0; i<CELL_SIZE; ++i)
13         data[i] = c.data[i];
14     }
15   }
16
17   enum State {WAIT, START, SEND};
18   State state = WAIT;
19   i = 1;
20   RTLSignals rtl;
21
22   void rtl_interface() {
23     if (state == WAIT) {
24       rtl.enb = 0;
25       if (rtl.clav) {
26         state = START;
27       }
28     } else if (state == START) {
29       assert (rtl.clav);
30       if (data_available) {
31         rtl.enb = 1;
32         rtl.soc = 1;
33         rtl.data = data[0];
34         state = SEND;
35       } else {
36         rtl.enb = 0;
37       }
38     } else if (state == SEND) {
39       rtl.enb = 1;
40       rtl.soc = 0;
41       rtl.data = data[i];
42       ++i;
43       if (i == CELL_SIZE) {
44         i = 1;
45         state = WAIT;
46         data_available = false;
47         notify(empty);
48       }
49     }
50   }
51 };
```

Fig. 8.3 TLM-to-RTL transactor relevant implementation pseudocode

Transactors Figure 8.3 shows the relevant implementation parts of the input transactor (TLM-to-RTL). Essentially, the transactor has two functions: a TLM input function to receive a cell and a RTL interface function to send it out. The TLM input function receives a cell and stores it internally. It will block (Line 8) in case a cell has already been received but not send out (*data_available=true*). The RTL interface function is sampled at every RTL clock cycle. It mimics the behavior of an ATM master device. Therefore, the transactor keeps track of the internal protocol status (Line 18). It starts in *status=WAIT* until the clav signal is observed. Then (*status=START*) it initiates data transfer once a TLM cell is available. In the next clock cycles (all *status=SEND*), the remaining data bytes are transferred (Line 41). The variable *i* tracks the next byte to send. Once the transfer is finished, the internal protocol data (status and i) are reset and the TLM input function is unblocked (Lines 44–47), i.e. can receive the next cell. Asserts are used in the transactor to detect illegal signal combinations (e.g. clav is set to low before data is sent). The implementation of the output transactor (RTL-to-TLM) is similar to the input transactor, and therefore omitted.

8.1.2 *Static Analysis of Transactors*

This section describes our static analysis based on symbolic execution to extract the underlying transactor protocol between TLM and RTL as FSMs.

8.1.2.1 Symbolic Execution

Starting with the initial symbolic execution state, the transactor function is repeatedly executed, until all reachable states are explored. Figure 8.4 shows the complete symbolic state space of the transactor. In total we have 7 distinct states ($\{S0, \ldots, S6\}$), which make up the state space, and 6 additionally observed states ($\{*S7, \ldots, *S12\}$) that are equivalent to one of the first 7 and thus are discarded.

The initial execution state S0 is defined as follows: RTL signals and TLM input are externally provided and therefore initialized with symbolic values. The protocol data (*status* and index *i*) are initialized with concrete values. Beside the variable environment, a symbolic execution state has a path condition (*PC*, initialized to *True*, i.e. no constraints in S0). Constraints are added when executing an *assert(C)*, i.e. $PC = PC \wedge C$, or branch with symbolic condition C, e.g. starting from S0 execution will fork due to symbolic signal *clav* resulting in the states S1 and S6 (PC updated as $PC = PC \wedge C$ and $PC = PC \wedge \neg C$, respectively). Between subsequent executions protocol data is preserved, whereas the RTL signal values are reset to

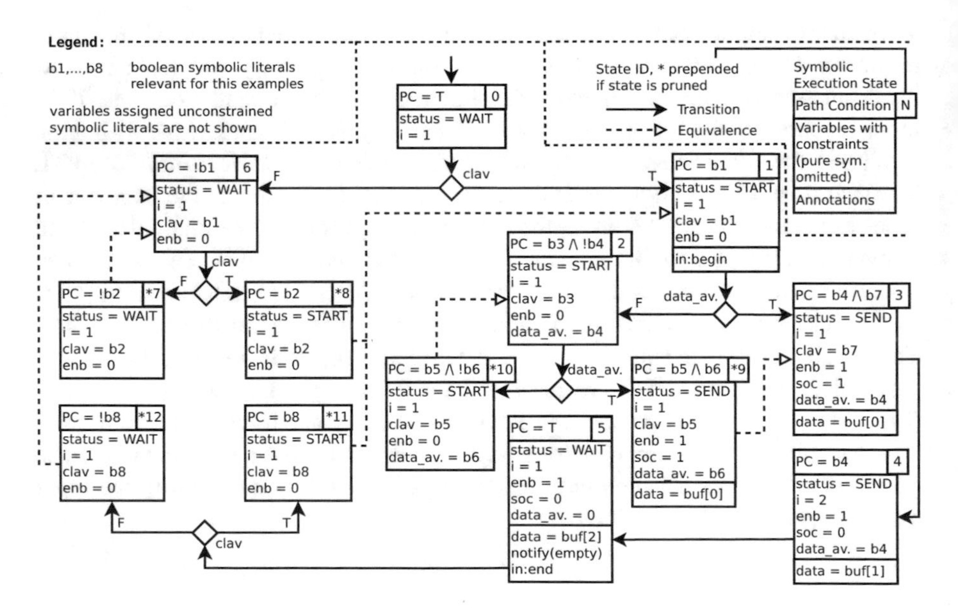

Fig. 8.4 Input transactor symbolic execution state space for the RTL interface function for the input transactor in Fig. 8.3, path conditions have been optimized by unused constraint elimination

fresh symbolic values each time (as they can be modified in each clock cycle). The TLM input data is preserved between executions where the TLM input function cannot be called. This is the case for S3 and S4, since data_available is set to *true*, which will block the input function. This information is collected by performing a preliminary static analysis on the input function. S5 unblocks the input function again by setting *data_available* to *false* and notifying the awaited event *empty*. We annotate the input related TLM events, i.e. *in:b* and *in:e*, to S1 and S5, respectively. The reason is that the execution of S1 is the first one to access input related data, and S5 the last one. Furthermore, we annotate the data mapping between the internal buffer *buf* and the RTL *data* signal in S3, S4, and S5. Combining it with the data mapping of the input function, there exists a mapping between the RTL *data* signal and TLM input cell at different states.

8.1.2.2 FSM Construction

The symbolic state space is transformed into an FSM by abstracting away all data except the RTL signal values. The FSM contains the same states and edges as the symbolic state space, but the edges are annotated with RTL signal values. For example starting from S0 either S1 (with $clav = 1$ and $enb = 0$) or S6 (with $clav = 0$ and $enb = 0$) can be reached. Please note that the value of clav is constrained to 1 (S1) and 0 (S6) due to the path condition. The other signal values are unconstrained and thus not mentioned. Signal combinations which are not covered, e.g. $clav = 0$ and $enb = 1$ in S0, are invalid based on the protocol. By repeating this process for all other edges, the transactor FSM is obtained. The input transactor FSM is shown in Fig. 8.5. We additionally preserve the TLM event annotations and data mappings in the FSM states. The output transactor FSM shown in Fig. 8.6 can be obtained similarly.

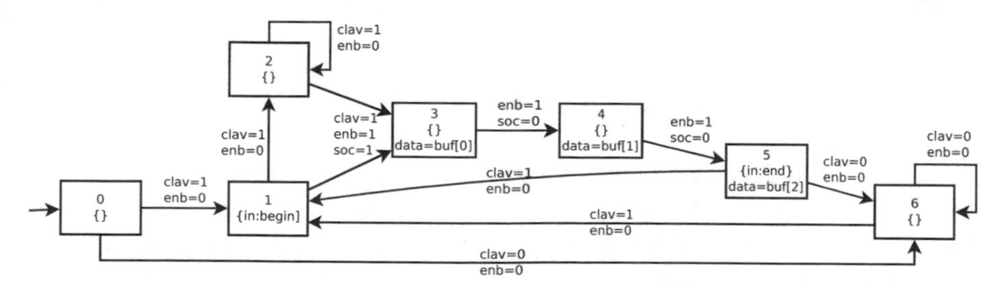

Fig. 8.5 FSM for the TLM-to-RTL transactor (input)

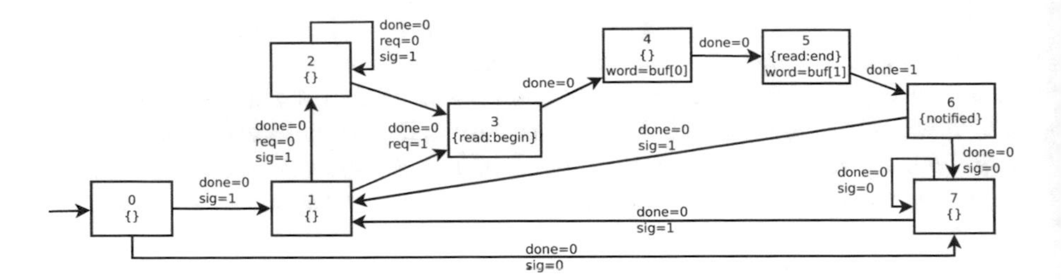

Fig. 8.6 FSM for the RTL-to-TLM transactor (output)

8.1.3 *Property Refinement*

This section outlines the refinement process based on the extracted FSMs. Before going into details, we briefly describe the used property language as well as the actual TLM properties for refinement.

8.1.3.1 Property Specification Language

For property specification, we consider the *simple subset* of PSL with TLM extensions similar to [55, 205]. For TLM properties, we use TLM events as atomic boolean propositions. For every function f we define the events *f:begin* and *f:end* (abbreviated as *f:b* and *f:e*). Arguments can be captured to reason about the data, e.g. *in(x):e* denotes that the *in* function has finished with x being passed as parameter. Finally, *e:notified* denotes that event *e* has been notified. Boolean C++ expressions can be embedded to reason about data relations. We use SEREs (sequentialized regular expressions) to specify RTL traces (i.e. signal sequences sampled at clock cycles).

8.1.3.2 TLM Properties for Refinement

We consider two TLM properties P1 and P2, shown in upper part of Table 8.1. P1 specifies that after a DMA read a DMA event notification needs to happen before the next DMA read is started. This ensures that data is available before it is attempted to read. P2 checks that no packets are lost. It captures the input packet in the variable x and when the input is completed, either the next or second next DMA read will read x. The *next_event*(α, ϕ) operator requires that ϕ is true at the next occurrence of α. The $CEQ(y, x)$ predicate returns true when the DMA result and input cell are equal. The predicate can be expressed in C++ syntax as shown in Table 8.1 (upper part).

Table 8.1 Mapping of TLM-to-RTL properties with refinement steps

<table>
<tr><td>CEQ(y,x)</td><td>(y.ptr[0] & 0xff == x.data[0]) ∧ (y.ptr[0] » 8 == x.data[1]) ∧ (y.ptr[1] & 0xff == x.data[2])</td></tr>
<tr><td>P1</td><td>always(read:e -> notify before read:b)</td></tr>
<tr><td>P2</td><td>always(in(x):e -> next_event(read(y):e, CEQ(y,x) | next(next_event(read(z):e, CEQ(z,x)))))</td></tr>
<tr><td>RTL(notified)</td><td>done</td></tr>
<tr><td>_H1</td><td>(!done&&!sig | done ; !done&&!sig) | done ; !done&&sig ; !done&&req</td></tr>
<tr><td>_H2</td><td>(!done&&!sig | done ; !done&&sig) | !done&&!req&&sig ; !done&&!req&&sig</td></tr>
<tr><td>RTL(read:b)</td><td>_H1 | _H2 ; !done&&req</td></tr>
<tr><td>RTL(read:e)</td><td>!done&&sig | !done&&sig&&!req ; !done&&req ; !done ; !done</td></tr>
<tr><td>RTL(in(x):e)</td><td>(enb&&clav&&soc):NEW(x_0:=data) ; (enb&&!soc):NEW(x_1:=data) ; (enb&&!soc):NEW(x_2:=data)</td></tr>
<tr><td>_S4</td><td>(done | !done&&!sig ; !done&&sig) | !done&&!req&&sig ; !done&&req ; !done</td></tr>
<tr><td>RTL(read(y):e)</td><td>_S4:NEW(y_0:=word) ; !done:NEW(y_1:=word)</td></tr>
<tr><td>RTL(CEQ(y,x))</td><td>(y_0 & 0xff == x_0) ∧ (y_0 » 8 == x_1) ∧ (y_1 & 0xff == x_2)</td></tr>
<tr><td>RTL(P_1)</td><td>always(RTL(read:e) |-> RTL(notify) before RTL(read:b))</td></tr>
<tr><td>RTL(P_2)</td><td>always(RTL(in(x):e) |-> next_event(RTL(read(y):e), RTL(CEQ(y,x))) | next(next_event(RTL(read(z):e), RTL(CEQ(z,x)))))</td></tr>
</table>

Upper half shows the TLM properties and equality predicate CEQ, middle half shows mapping of TLM events to RTL signal sequences, and lower half shows the refined results.

8.1.3.3 Refinement Process

Our refinement works recursively. The operators *always*, *next*, *before*, |, and *next_event*(α, ϕ) are preserved. The implication L `->` R is mapped to L `|->` R, i.e. first L needs to be matched and matching of R starts at the clock cycle L has been matched at. This is a valid refinement thanks to the monotonic advancement of time required by the simple subset.

A TLM event E is mapped to a sequence of RTL signals (i.e. traces) by following back all edges starting from FSM state S, event E is associated with, and collecting the signal assignments until the resulting traces T uniquely identify S, i.e. starting from any state Q in the FSM and following any $t \in T$, only state S can be reached (otherwise there has been an invalid transition). We call T the distinguishing traces of state S, and can straightforwardly express T as a SERE by sequencing (operator `;`) all signal assignments within a trace and then union all traces (operator |). Table 8.1 (middle part) shows the results for the relevant TLM events. For example *notified* (associated with state S6 in output FSM) is uniquely identified by *done* (i.e.

done = 1), as S5 is the only state where *done* is a valid signal assignment. On the other hand to identify S5 (associated with *read:e*) going back one step is not enough, as many states can be reached with *!done*. In fact, T for S5 contains two traces of length 4.

Function call events capturing input/output arguments require an extended refinement procedure. It is not enough to just visit the FSM state associated with the function begin/end, but also go through all states where a data mapping to the arguments exists. The event *in(x):end* requires to visit the state sequence *[S3, S4, S5]* in the input FSM. Therefore, we generate T for the first state in the sequence and extend it with signal assignments to reach the subsequent states by following along the FSM edges. At each state we capture the current data signals in a new local variable for S3, e.g. *NEW*(x_0*:=data)* (see [170]). Based on the data mapping from the FSMs, we obtain that RTL signal *data* in S3 (input FSM) is mapped to *buf[0]* and following along the data mapping of the TLM input function, that *buf[0]* maps to *c.data[0]* of the input cell. Similarly, we obtain mappings for the output side. With the captured data signals and mappings, we can automatically refine the *CEQ(y,x)* predicate (Table 8.1 lower part).

8.1.4 Discussion and Future Work

Our proposed *fully automated* TLM-to-RTL property refinement approach is still a work-in-progress. The monotonic advancement of time through the formula allows for an intuitive recursive refinement algorithm, where evaluation of subformulas starts at the clock cycle the previous one finished at. Although the feasibility of our refinement approach has been demonstrated on a realistic case study with two representative TLM properties, a formal characterization of supported properties is still to be defined in a future work. One limitation is that we assume the next (or SEREs in general) operator is not used to specify an absolute order of TLM events. The reason is, that it is unclear how to automatically refine it (especially when combined with other operators), as there can be (arbitrary) many clock cycles between TLM events at RTL. Moreover, it has been suggested that using the next operator in this way at TLM is not well suited (up to the point of being unsound) due to its restrictiveness [170]. Rather, the *before* and *next_event* operators, which specify precedence, should be used. However, using a single next operator in combination with -> and *next_event* makes sense to avoid matching the subformula at the current instant (e.g. as used in our property P2 in Sect. 8.1.3.3 to avoid matching the same *read*). This issue needs further investigation.

For property refinement, we compute distinguishing traces for FSM states. However, this is not possible for FSMs that have multiple state cycles, which accept the same input sequence. For a protocol description such an FSM implies that it is not possible to detect the protocol status by just observing the IO signals, i.e. there is no synchronization or handshaking in place. While we believe it is a reasonable assumption to have distinguishing states, our refinement algorithm can be extended to handle non-distinguishing states as well by: (1) accessing internal variables of the model or (2) generating additional logic between model and monitor to simulate the FSM (can be done automatically). We plan to evaluate these extensions in future work.

8.2 Automated RTL-to-TLM Fault Correspondence Analysis

Ensuring the functional safety of electronic systems becomes one of the most important issues nowadays, as these systems are being more and more deeply integrated into our lives. Even if a system can be proved to perform its intended functionality correctly, failures are still possible due to Hardware (HW) faults caused e.g. by radiation or aging. The risk of such faults is rapidly increasing with the raising complexity of design and technology scaling. To evaluate and facilitate the development of safety measures, fault injection is a widely accepted approach, which is also recommended in different functional safety norms such as IEC 61508 and ISO 26262.

Traditional HW fault injection approaches operate on low levels of abstraction such as gate level or *Register-Transfer Level* (RTL). While physical faults can be quite accurately modeled at these abstraction levels, the slow simulation speed becomes a major bottleneck for modern systems. This used to be a problem for Software (SW) development and system verification as well, until the emergence of *SystemC Virtual Prototypes* (VPs). VPs are basically full functional SW models of HW abstracting away micro-architectural details. This higher level of abstraction, often termed as *Transaction-Level Modeling* (TLM) [113], allows significantly faster simulation compared to RTL. Therefore, safety evaluation using VP-based fault injection, envisioned as error effect simulation in [160], is a very promising direction.

Fault injection techniques for SystemC have attracted a large number of work, see e.g. [137, 153, 168, 193]. They form a technical basis for error effect simulation, but do not focus on the core problem: the higher level of abstraction of TLM poses a big challenge in error modeling. Please note that we use the term "fault" for RTL and "error" for TLM due to the fact that TLM is just a modeling abstraction. A high-level error model, if not carefully designed, would yield significantly different simulation results compared to a low-level fault model [35, 140]. This would make the safety evaluation results using VPs to become misleading. Unfortunately, deriving an accurate high-level error model is very difficult [35, 140].

This book examines the idea of a novel cross-level fault correspondence analysis to aid the design of such error model. The prerequisite of our analysis is the availability of an RTL model and its corresponding TLM model. Then, for a given RTL fault model, our analysis automatically identifies for an RTL fault a set of candidates for error injection in the TLM model. These candidates are potentially *equivalent* to the RTL fault, in the sense that the error-injected TLM model would produce the same failure as the fault-injected RTL model.

The core idea of this RTL-to-TLM fault correspondence analysis is inspired by the concept of *formal fault localization* (see [69] for C and [133] for SystemC TLM). These approaches automatically identify candidates for modifications in the model under verification, which would make a given set of failing testcases to become passing testcases. Such a candidate potentially points to the location of the bug that causes a failure. Our analysis is dual to this: Given successful TLM simulations,[1] we search for locations to inject an error to produce the same failure observed in faulty RTL simulations. Hence, we can apply the same set of techniques: instrumenting the TLM model to include non-deterministic errors and leveraging existing TLM model checkers to compute the candidates.

We also present first results of a case study on the *Interrupt Controller for Multiple Processors* (IRQMP) of the SoCRocket VP, which is being used by the European Space Agency [189], to demonstrate the feasibility of the analysis. In particular, we implement the cross-level analysis on top of the recent symbolic simulation approach for SystemC [134, 135] and apply our analysis to find TLM error injection candidates for transient bit flips at RTL. To the best of our knowledge, it is the first time such results on fault correspondence between TLM and RTL are reported.

As mentioned earlier, a large number of fault injection techniques for SystemC TLM models exists. These approaches assume some TLM error models without qualitative or quantitative assessment of their correspondence to RTL faults. Beltrame et al. [14] includes a comprehensive list of TLM errors that might result from RTL bit flips. The book also provides a set of guidelines on how to (manually) derive corresponding TLM errors, whereas our analysis is automated.

Another line of work is mutation analysis for SystemC TLM models [22, 143, 191]. At the heart of any mutation analysis is also an error model. However, the purpose of such model is to mimic common design errors, but not HW faults caused by external impact.

The most close work to ours is the one in [25], which proposes an automatic RTL-to-TLM transformation to speed up RTL fault simulation. The transformation produces an equivalent TLM model from each fault-injected RTL model. The obtained results can be mapped back to RTL. However, this approach relies on a particular RTL-to-TLM transformation. Such transformation might not provide the best possible speed-up compared to hand-crafted TLM models. In contrast to our analysis, this approach is not applicable when the corresponding TLM model already exists.

[1] A successful TLM simulation produces the same output as RTL simulation under the same inputs.

8.2.1 RTL-to-TLM Fault Correspondence Analysis

Our proposed RTL-to-TLM fault correspondence analysis allows to find errors in a TLM model that correspond to faults in the RTL model. We assume that the RTL and TLM model are functionally equivalent, i.e. they produce the same outputs when given the same inputs. Furthermore, we assume that a set of (representative) inputs, e.g. in the form of testcases, is available for the RTL or TLM (since both use the same inputs) model. These inputs should preferably cover a large set of functionality of the design. Similarly, we assume that a set of fault injection locations is available for the RTL model. Otherwise, the fault injection locations can be obtained by tracing the execution of the RTL model, e.g. using an observer class, based on the available testcases. Please note that a fault injection location consists of three pieces of information: (1) a source line, (2) an injection time, i.e. a number that denotes which execution of this source line should be fault injected, and (3) the bit position which shall be flipped. The reason for information (2) is that we consider transient faults.

8.2.1.1 Correspondence Analysis Overview and Algorithm

Figure 8.7 shows an overview of our analysis approach. A corresponding algorithm is shown in Algorithm 3. It computes a set of corresponding error candidates on TLM level for every injected fault on RTL. The algorithm considers every fault injection location L on the RTL model. First we construct the faulty RTL model with respect to L (Line 4). In our implementation, we attach a fault injection class

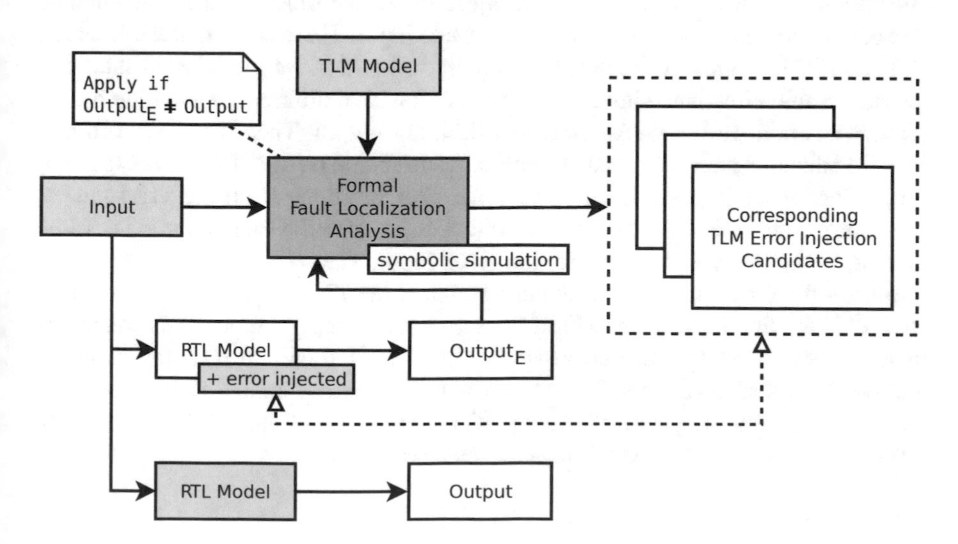

Fig. 8.7 Fault correspondence analysis overview

Algorithm 3: Fault correspondence analysis

```
1  result ← ∅ /* Mapping from RTL fault injection location to set
       of corresponding TLM error injection candidates            */
2  for L ∈ RTL-fault-injection-locations do
       /* Start with empty result set for L                        */
3      result[L] ← ∅
4      RTL-Model_E ← RTL-Model with L injected
5      for input ∈ testcases do
           /* Check if RTL fault L has observable effect           */
6          output ← simulate(RTL-Model, input)
7          output_E ← simulate(RTL-Model_E, input)
8          if output ≠ output_E then
9              candidates ← fault-localization-analysis(    TLM-Model,
                input, output_E)
10             if result[L] = ∅ then
                   /* First set of candidates                      */
11                 result[L] ← candidates
12             else
                   /* Combine with existing set                    */
13                 result[L] ← result[L] ∩ candidates
14             if result[L] = ∅ then
                   /* No corresponding TLM error for L found, check
                      next RTL fault                               */
15                 break
```

as observer to our RTL model that can inject an error at runtime. Then we simulate the correct and faulty RTL model with the same input. There are two possible cases: (1) Both RTL models produce the same output. In this case the injected fault had no observable effect and simulation is repeated with a different input (Line 5). (2) Otherwise, both RTL models produce a different output. Then we apply a formal fault localization analysis based on symbolic simulation on the TLM model (Line 9) to produce a set of possible corresponding error locations C on the TLM model. Essentially, all these reported error locations C produce the same failure, i.e. same output, as the faulty RTL model for the given input. The set C is integrated into the result set for L by computing a set intersection (Line 13)—or simple assignment in case C is the first result (Line 11). If the result set is empty, our analysis concludes that no corresponding TLM error can be found for the current RTL fault and we consider the next RTL fault. The reason is that a corresponding TLM error must produce the same output as the faulty RTL model for *all* inputs. Otherwise, the result set is not empty, we continue with the next input.

8.2.1.2 Example

Interrupt Controller for Multiple Processors (IRQMP) We first briefly describe the interrupt controller IRQMP of the SoCRocket VP that is used in this example and later in the case study. The IRQMP is an interrupt controller from the SoCRocket VP supporting up to 16 processors [189]. It consists of a register file, several input and output wires, and an APB Slave bus interface for register access. The register file contains shared and processor-specific registers. Every register has a bit width of 32. Each bit naturally represents an interrupt line.

The IRQMP supports incoming interrupts (using the *irq_in* wire or *force* register) numbered from 1 to 31 (interrupt line 0 is reserved/unused). Lines 15:1 are regular interrupts and lines 31:16 are extended interrupts. In regular operation mode, IRQMP ignores all incoming extended interrupts. The *irq_req* and *irq_ack* wires are connected with every processor and allow to send interrupt requests and receive acknowledgements.

The functionality of the IRQMP is to process incoming interrupts by applying masking and prioritization of interrupts for every processor. Prioritization of multiple available interrupts is resolved using the *level* register. A high (low) bit in the *level* register defines a high (low) priority for the corresponding interrupt line. On the same priority level, interrupt with larger line number is higher prioritized. See the specification [68] of the IRQMP for more details.

Method Overview As an example, consider a bit flip fault in the RTL model of the IRQMP when initially configuring the mask register using a bus transfer operation as fault injection location L. Furthermore, consider three test scenarios with different inputs:

1. Send interrupt using the *irq_in* (incoming interrupt) wire
2. Send interrupt using the *force* register
3. Send the same interrupt twice using the *irq_in* wire

All of these inputs result in different outputs for the RTL model with and without injection of the fault L. In particular no interrupt is generated by the interrupt controller for the masked interrupt line. On the TLM model, our analysis identifies different candidates for corresponding errors. In particular, it identifies different transient bit flip errors during computations as well as wire/bus transfer that lead to the same observable behavior.

The results are summarized in Table 8.2. For the first and second input we obtain multiple possible error locations in the TLM model. By sending the same interrupt twice (third input) to the interrupt controller, the effects of a transient computation error and non-related transfer error are eliminated. By computing the intersection of all possible error locations for these inputs, the corresponding TLM error is obtained—bus transfer error during configuration of the mask register.

Fault Correspondence To further illustrate this fault injection example, Figs. 8.8 and 8.9 show relevant code of the IRQMP for configuring the *mask* register at RTL and TLM, respectively. The source line where the fault has been injected at RTL and

Table 8.2 Corresponding TLM error injection candidates (corresponding error highlighted)

TLM error injection location	Inputs 1	2	3
Bus transfer			
Mask register configuration	X	X	X
Force register configuration		X	
Wire transfer			
Incoming interrupt wire	X		
Computation			
Prioritization Logic 1	X	X	
Prioritization Logic 2	X	X	

```
1  -- combinatorial VHDL process, which essentially contains the whole logic of the
       IRQMP - sensitive on the reset signal, update of internal signals as well as
       incoming bus and interrupt signals
2  comb : process(...)
3    -- send out current internal signal values and compute new values which will
         overwrite the current values in the next clock cycle
4    variable v : reg_type; -- registers to store regular interrupt lines for local
         computation
5    variable v2 : ereg_type; -- registers to store extended interrupt lines for local
         computation
6    begin
7      -- ... prioritize interrupts, register read ...
8
9      -- register write
10     if ((apbi.psel(pindex) and apbi.penable and apbi.pwrite) = '1' and
11       (irqmap = 0 or apbi.paddr(9) = '0')) then -- essentially, check that bus is
            enabled and used in write mode
12       case apbi.paddr(7 downto 6) is -- decode target register
13       -- ...
14       when "01" => -- write to processor specific mask register
15         for i in 0 to ncpu-1 loop -- iterate over all processors
16           if i = conv_integer( apbi.paddr(5 downto 2)) then -- decode and check
                target processor
17             v.imask(i) := apbi.pwdata(15 downto 1); -- write to mask of processor i,
                 RTL fault injected here
18             if eirq /= 0 then -- check if extended interrupts are also handled
19               v2.imask(i) := apbi.pwdata(31 downto 16); -- in this case also update
                   the extended interrupt lines of the mask register
20             end if;
21           end if;
22         end loop;
23       -- ...
24       end case;
25     end if;
26
27     -- ... register new interrupts, interrupt acknowledge, reset ...
28 end process;
```

Fig. 8.8 Example IRQMP code excerpt in VHDL showing register configuration, to illustrate fault injection on RTL (line with fault injection highlighted)

```
1  // generic (using templates) and
       re-usable TLM register class,
       the inherited class provides the
       basic interface and some default
       implementation
2  template<typename DATA_TYPE>
3  class sr_register : public
       sc_register_b<DATA_TYPE> {
4    //...
5  public:
6    void bus_write(DATA_TYPE i) {
7      // callbacks notify observers
          about register access
          directly without context
          switches
8      raise_callback(SR_PRE_WRITE);
9      // corresponding TLM error
          location - update the
          internal register value, the
          write mask allows selective
          updates
10     this->write(i & m_write_mask);
11     // the IRQMP will trigger
          interrupt re-computation
          after the mask register is
          updated
12     raise_callback(SR_POST_WRITE);
13   }
14   //...
15 }
16
17 // a register bank groups multiple
       registers - similarly, this is a
       generic implementation
18 template<typename ADDR_TYPE, typename
       DATA_TYPE>
19 class sr_register_bank : public
       sc_register_bank<ADDR_TYPE,
       DATA_TYPE> {
20   typedef typename
          std::map<ADDR_TYPE,
          sr_register<DATA_TYPE> *>
          register_map_t;
21   register_map_t m_register; // use a
          mapping (address to register)
          to store registers
22   //...
23 public:
24   // bus read/write transactions
          matching a register address are
          automatically redirected to
          this class
25   bool bus_write(ADDR_TYPE offset,
          DATA_TYPE val) {
26     sr_register<DATA_TYPE> *reg =
          get_sr_register(offset); //
          retrieve register for the
          given (bus) address
27     if(reg) {
28       reg->bus_write(val); // update
            register value
29     }
30     return true;
31   }
32   //...
33 }
34
35 // register bank used as member
       variable in the IRQMP
36 sr_register_bank<unsigned int,
       unsigned int> r;
```

Fig. 8.9 TLM code for register configuration, showing a corresponding TLM error (line highlighted) for Fig. 8.8

the corresponding error location at TLM are highlighted. The RTL code is available in VHDL, and the TLM code in SystemC.

The RTL code stores internal signals for registers separated for regular (lines 15:1, *reg* type) and extended interrupts (lines 31:16, *ereg* type). The processing logic of the IRQMP is available in the combinatorial process *comb*, shown in Fig. 8.8. Essentially, it contains the whole logic of the RTL model and is triggered whenever some input or internal signal changes. It is responsible for interrupt prioritization, processing of register read/write requests and interrupt acknowledgements, and writing output signals. Internal signal values are updated in a separate process at every clock cycle. In particular Fig. 8.8 shows processing code for a bus write request to the CPU specific *mask* register for normal (Line 17) and extended interrupts (Line 19). The target register and processor are encoded in the bus address signal, they are decoded in Lines 22 and 16, respectively. In this example a fault is injected in Line 17 during *mask* register configuration.

The TLM implementation, shown in Fig. 8.9, of the IRQMP keeps a register bank (Line 36), which essentially contains a mapping of an address value to a register (Line 21). A bus write transaction will call the *bus_write* function of the register bank (Lines 25–31). The function will retrieve the target register directly based on the write address in Line 26 and dispatch the write access to the register class in Line 28. The register will finally update its internal value in Line 10, which is the corresponding TLM error. Before and after the update, callback functions are used to notify the IRQMP and update its internal state directly without any context switches. In this case the main processing thread of the IRQMP will be notified to re-compute outgoing interrupts. Please note that the register bank implementation is not specific to the IRQMP but a generic and re-usable implementation. Furthermore, the TLM model can update the whole register, while the RTL model only updates bits 15:1 when extended interrupts are ignored. This is not a problem, since the prioritization logic of the TLM model simply ignores the extended interrupt lines in this case and therefore produces the same failure (when the corresponding error is injected) as the faulty RTL model.

8.2.2 Formal Fault Localization Analysis

This section describes our formal fault localization analysis, to obtain candidates for corresponding TLM errors, in more detail. We reduce this problem to a verification problem of assertion violations by encoding error injection selection non-deterministically and adding appropriate constrains to prune invalid solutions. Then we employ symbolic simulation for an efficient exhaustive exploration to find all possible solutions. In general different formal verification techniques besides symbolic simulation could also be used to find solutions.

Figure 8.10 shows an overview of our approach. The analysis requires the TLM model as well as the input and output of the faulty RTL model (marked gray in Fig. 8.10). We assume that the TLM model is available in the XIVL format to apply formal analysis techniques. The TLM model contains annotations for a fine grained selection of instruction where error injection can take place (see 1 in Fig. 8.10).

First a testbench is generated by using the input and output to construct an input driver and result monitor, respectively (see 3 in Fig. 8.10). The testbench is also available in the XIVL format and contains the simulation entrypoint—the main function—which is responsible to setup all components. The result monitor contains the assertions which constrain valid solution.

Then the TLM model and testbench are automatically combined to a complete TLM model. During this process, the TLM model will be instrumented with symbolic error injection logic to non-deterministically select an error injection location for a transient one bit error (see 3 in Fig. 8.10). The annotations on the TLM model are used to guide the instrumentation.

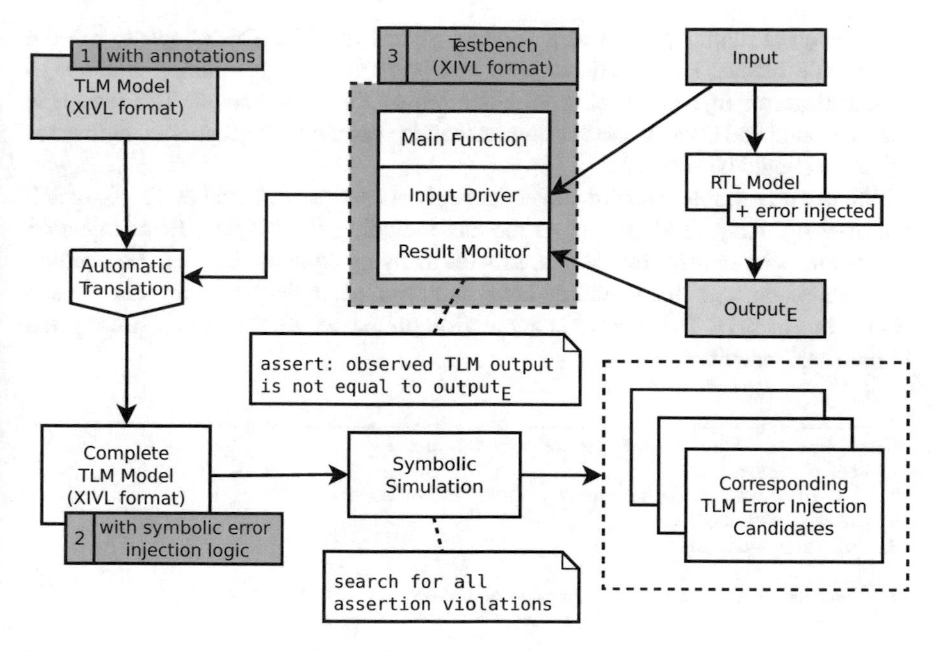

Fig. 8.10 Formal fault localization analysis overview

Finally, the complete TLM model is passed to our symbolic simulation engine for a formal analysis. Based on the symbolic error injection logic instrumented into the TLM model, the symbolic simulation will find and report all concrete error injection locations that will cause the TLM model to produce the same failure, i.e. same output, as the faulty RTL model for the given input.

In the following we will discuss (1) annotations, (2) symbolic error injection logic, and (3) testbench encoding in more detail.

8.2.2.1 Annotations

We use annotations in the TLM model for a fine grained control of error injection. Assignment instructions are annotated to denote, that error injection can take place. During instrumentation, every annotated assignment will be modified to either stay unchanged or toggle a bit flip in the result—based on a non-deterministic choice during analysis (at runtime). For convenience, we also support function annotations. These will be propagated to all assignments in the function. Using annotations is a flexible approach to control error injection more precisely. These annotations can happen manually, or using a static analysis that modifies the code automatically, e.g. based on a specification provided by the user. This ensure that meaningful locations for error injection are reported—otherwise the initial state, or the output values would be modified. Furthermore, injection can be selectively activated and deactivated at runtime, based on a boolean global variable. This allows to run the same code blocks, e.g. functions, with and without error injection. We use it to deactivate error injection during initialization of the TLM model. The reason is that the same code is also used during testbench specific configuration, where error injection should be allowed.

We use a re-usable modeling layer for registers and wires based on [135], which are used by many TLM peripheral models including the IRQMP. Error injection in bus and wire transfer operations, as well as many computations can be handled by injecting errors in the modeling layer itself. An example for the register class is shown in Fig. 8.11. The *injectable* annotation on the *set_bit* function is propagated to both assignments.

```
1  // XIVL implementation of the TLM register class
2  struct Register {
3    // write mask allows selective updates on bus access
4    uint32_t value;
5    uint32_t write_mask;
6
7    bool get_bit(Register *this, uint32_t index) {
8      return this->value & (1 << index);
9    }
10
11   @injectable
12   void set_bit(Register *this, uint32_t index, bool value) {
13     // injectable annotation is automatically propagated to both assignments
14     if (value) {
15       // set bit
16       this->value |= 1 << index;
17     } else {
18       // clear bit
19       this->value &= ~(1 << index);
20     }
21   }
22
23   uint32_t read(Register *this) {
24     return this->value;
25   }
26
27   void write(Register *this, uint32_t value) {
28     // errors can be injected at assignments marked with @injectable
29     @injectable this->value = value;
30   }
31
32   void bus_write(Register *this, uint32_t value) {
33     // ... PRE/POST write callback handling omitted in this example ...
34     @injectable this->value = value & write_mask;
35   }
36 }
```

Fig. 8.11 Excerpt of an annotated TLM register class in XIVL

```
1  bool active = false; // dynamically toggle error injection
2  void deactivate_injection() { active = false; }
3  void activate_injection() { active = true; }
4  // variables store injection choices for inspection and ensure that only single
       error is injected
5  int id = 0;
6  int location = -1;
7  int condition = ?(int);
8  int bit = -1;
9
10 int64_t flip_single_bitvector_bit(int64_t val, uint8_t bitwidth) {
11   // non-deterministically choose a single bit in *bitwidth* to flip
12   uint8_t x = ?(uint8_t);
13   assume (x >= 0 && x < bitwidth);
14   val = val ^ (1 << x); // perform the actual bit flip
15   bit = x; // store choice for later inspection
16   return val; // result is symbolic due to non-deterministic choice
17 }
18
19 int64_t inject_bitvector_error(int64_t val, uint8_t bitwidth) {
20   id += 1; // unique id ensures only a single error is injected
21   if ((condition == id) && active) {
22     // record the injection choice for later inspection
23     @track "one bit error injection";
24     location = id;
25     // perform a non-deterministic bit flip
26     val = flip_single_bitvector_bit(val, bitwidth);
27   }
28   return val;
29 }
30
31 bool inject_bool_error(bool val) { // similar to "inject_bitvector_error"
32   id += 1;
33   if ((condition == id) && active) {
34     @track "boolean error injection";
35     location = id;
36     val = !val;
37   }
38   return val;
39 }
```

Fig. 8.12 Encoding details for transient one bit error injection

8.2.2.2 Symbolic Error Injection Logic

We integrate a set of global variables and functions, shown in Fig. 8.12, into the complete TLM model for symbolic error injection. In particular the functions *inject_bitvector_error* and *inject_bool_error* are used from within the TLM model. Consider again the example in Fig. 8.11. For example, the assignment *@injectable this->value = value;* in the *write* function in Line 29 is transformed to *this->value = inject_bitvector_error(value, 32);* when annotations are resolved during instrumentation.

The *inject_bitvector_error* function (defined in Fig. 8.12) expects two arguments, an integer and its bitwidth. The bitwidth argument denotes the range of bits from which one is selected non-deterministically for flipping. The bitwidth argument is automatically generated based on static type informations during the instrumentation process, which rewrites the annotations. By using a 64 bit integer type as argument and return value, we also automatically support all integer types with smaller bitwidth. For this case study, 32 bit values are sufficient. The *inject_bool_error* function in principle works analogously.

For convenience we use short names in Fig. 8.12 for the global variables, in the TLM model they have a unique name prefix to avoid name clashes with existing code. The boolean *active* variable allows to selectively toggle error injection and thus control it more precisely. It is accessed by means of the *activate/deactivate_injection* functions defined in Lines 3 and 2, respectively.

The *condition* variable is initialized with a symbolic integer value in Line 7 and together with the *id* variable allows non-deterministic selection of an injection location. This works as follows: Consider an invocation of *inject_bitvector_error* and assume that *active* is true. If no error has been injected yet, then both branch directions in Line 21 are feasible, i.e. $condition = id$ can evaluate to *true* and *false*. Therefore, symbolic execution will split into two independent paths S_T and S_F, respectively, and explore both branch directions. This will update the path conditions of S_T and S_F with $condition = id$ and $condition \neq id$, respectively. Please note, that *id* is a concrete integer value in this case, e.g. the number 4. Since *id* is incremented on every call of *inject_bitvector_error*, the true branch of the *if* statement in Line 21 becomes infeasible for the S_T path and all its descendants. Therefore, at most a single error is injected on every execution path. The *location* variable stores a copy of the injection *id* for debugging purposes.

The *@track* instruction is specifically recognized by our symbolic simulation engine and records a snapshot of all instruction pointers of the callstack of the currently executed thread, i.e. essentially the currently executed instruction in the active thread and all of its called functions. This allows to pinpoint the error injection location for later inspection.

The function *flip_single_bitvector_bit* is used in Line 26 as a helper function to inject a single bit error into an integer variable. It creates and constrains a symbolic integer value to non-deterministically select a single bit in the range defined by the *bitwidth* argument (Lines 12–13). The global *bit* variable records the non-deterministic choice in Line 15 for later inspection (eventually the SMT solver will provide concrete values for non-deterministic choices). Based on the non-deterministic choice, the function performs the bit flip in Line 14 and returns the result. Please note that the result itself becomes a symbolic expression.

8.2.2.3 Testbench

This section provides more details on the testbench focusing on assertion generation to guide the formal analysis. For illustration purpose, we discuss a (simplified) concrete example testbench for the IRQMP. Essentially, the input specifies incoming interrupts for the IRQMP and the output is a prioritized list of interrupt requests generated by the IRQMP.

When sending the interrupt mask $0b110$ as input and injecting a fault in the RTL model that results in wrong prioritization, the output [2, 3] is observed instead of the expected output [3, 2]—since higher interrupt lines have higher priority. Based on the input and faulty output the testbench is constructed. The monitoring logic records the observed interrupts in an array *irq*. Furthermore, it keeps track of the number of received interrupts in the *num_irqs* variable. Finally, the monitor asserts that *((irq[0]* $\neq$ *2) || (irq[1]* $\neq$ *3) || (num_irqs* $\neq$ *(2))* holds at the end of simulation. Essentially, it asserts that the observed output for the TLM model is not equal to the output of the faulty RTL model. Thus, the symbolic simulation engine will search for all possible error inject locations, that violate the assertion, i.e. produce the same failure at TLM as the faulty RTL model.

As an optimization, to prune irrelevant search paths which cannot produce the output of the faulty RTL model, we place *assume* instructions in the monitor. For this example, we would assume that the first received interrupt is 2 and the second is 3. Furthermore, we would assume that *num_irqs* < 2. Then a simple *assert (false);* can be placed at the end of simulation. Using stepwise assumptions during symbolic simulation, instead of a single assert in the end, can significantly reduce the considered search space, by pruning irrelevant search paths early.

8.2.3 Case Study

We have evaluated our proposed fault correspondence analysis on the IRQMP model from the SoCRocket VP [189] as a case study. Our formal fault localization analysis is based on symbolic simulation approach of [134, 135]. In the following we report first experimental results for our proposed approach.

8.2.3.1 Experiments

All experiments were performed on an Intel 2.6 GHz machine with 16 GB RAM running Linux. We employ Z3 v4.4.1 as our SMT solver.

For the experiments we use a set of representative test scenarios to cover different functionality of the IRQMP. Every scenario is using different inputs. Incoming interrupts are sent from the testbench to the IRQMP via register access (using a bus transfer operation) or by writing to the *irq_in* wire. Furthermore, different priority levels are tested and resending of interrupt requests. Please note, all tests only use a single CPU and do not consider extended interrupts. The reason is that the RTL and TLM model are not functionally equivalent when using these features.

We use a set of representative fault injection locations for the RTL model. For every of these fault locations, a fault is injected in the RTL model for every test scenario. Table 8.3 shows a summary of our experimental results for injecting a bit flip during register configuration, interrupt prioritization computation as well as interrupt sending and acknowledgment. In particular, the following faults are injected:

1. A bit is flipped in the mask register, therefore the corresponding interrupt line is not processed or additional interrupt line activated;
2. A bit is flipped in the force register, therefore an additional interrupt needs to be processed or is omitted;
3. An incoming interrupt line is flipped, which has similar effect as the previous fault, but covers different functionality of the IRQMP;
4. A bit is flipped when resetting the incoming interrupt lines to zero;
5. The activation signal of the active interrupt acknowledgment wire is flipped, therefore an acknowledgment is missed;
6. A bit is flipped when sending the acknowledgment, this results in a wrong interrupt being acknowledged.
7. A bit is flipped when computing the interrupt prioritization, therefore a wrong interrupt is send or the order of two incoming interrupts is changed (e.g. flipping the second bit in 0b11 results in 0b01, thus interrupt line 1 is send first, even though line 2 has higher priority—since line 2 is still in the pending register, it will eventually be sent out too);
8. Similar to the previous one, but at different location, where interrupts are interpreted as number instead of lines, e.g. 0b11 is interpreted as number 3 instead of interrupt lines 1 and 2;

Essentially, Table 8.3 combines the analysis results for all test scenarios for every fault injection location and reports the average values over all analysis runs. The first column shows a description of the RTL fault. The second column shows the type of the fault: *BT=Bus Transfer*, *WT=Wire Transfer* and *Comp=Computation*. The third and fourth column show the average runtime (in seconds), which is further divided in total analysis and SMT time. The SMT time shows how much of the analysis time is spent with solver queries. The number of solver queries is reported

Table 8.3 Combined results of experiments (Runtimes in Seconds)

		Average runtime			SMT	TLM errors		
RTL fault location	Type	TOTAL	SMT	#Inject.	Queries	Max	Min	Res
(1) Mask register configuration	BT	285.86	184.18	162	30,992	6	1	1
(2) Force register configuration	BT	317.88	210.14	196	34,690	7	4	4
(3) Incoming interrupt wire	WT	317.80	207.18	166	35,057	6	2	2
(4) Incoming interrupt wire reset	WT	690.74	454.03	311	74,595	0	0	0
(5) Missing acknowledgement	WT	395.10	262.00	211	42,839	7	6	6
(6) Wrong acknowledgement	WT	363.81	241.43	211	41,502	1	1	1
(7) Prioritization Logic 1	Comp	209.41	154.01	127	21,856	12	4	4
(8) Prioritization Logic 2	Comp	373.88	242.48	201	40,538	9	0	0

in the column *SMT Queries*. The column *#Inject* shows how many (symbolic) errors have been injected during analysis of the TLM model. Please note that every error injection represents a non-deterministic bit flip. Finally, the column *TLM Errors* shows the maximum (column *Max*), minimum (column *Min*), and result (column *Res*) of error injection locations detected on the TLM model. The result denotes the number of TLM errors when computing the intersection of TLM errors for every test scenario. In other words the result column denotes the number of candidates for corresponding TLM errors found by our analysis.

For the RTL faults 4 and 8, no corresponding TLM errors are found (result column contains 0). The reason is that the TLM model is designed on a higher level of abstraction. For performance reasons callbacks are used that directly modify the internal model state without any delta cycles and context switches. Therefore, in case of RTL fault 4, a reset of the incoming interrupt lines is not necessary (and not available in the TLM model—setting the interrupt lines to a specific value is only applied once and not permanently until change). In the other case (RTL fault 8) it is due to a signal line where an error effect is delayed and propagated across a delta cycle. For both cases, it is possible that the RTL fault corresponds to multiple simultaneous TLM errors. For their localization, the symbolic error injection logic must be extended to support multiple errors. This extension is left for future work. Finally, SW-based symbolic fault injection techniques, such as [136, 166], can complement this work.

8.3 Summary

In this chapter we proposed two approaches that perform a correspondence analysis between TLM and RTL to improve the VP-based design flow.

The first approach is a *fully automated* TLM-to-RTL property refinement. It employs a static transactor analysis based on symbolic execution to reverse-engineer the executable transactors in order to create a formal specification of the underlying protocol as FSM. Then, TLM properties are refined by relating high-level TLM events with RTL signal combinations at different clock cycles based on the FSM. This approach helps to avoid the manual transformation of properties from TLM to RTL in the design flow which is error prone and time consuming.

The second approach we proposed is an RTL-to-TLM fault correspondence analysis to improve the quality of error effect simulation using VPs. We employ formal methods to identify TLM error injection candidates for transient bit flips at RTL. First experiments on the IRQMP from the SoCRocket VP demonstrated the applicability and effectiveness of our approach in finding a small set of candidates for corresponding TLM errors.

Chapter 9
Conclusion

In the last years the complexity of embedded devices has been increasing steadily with various conflicting requirements. On the one hand, IoT devices need to provide smart functions with a high performance including real-time computing capabilities, connectivity, and remote access as well as safety, security, and high reliability. At the same time they have to be cheap, work efficiently with an extremely small amount of memory and limited resources, and should further consume only a minimal amount of power to ensure a very long runtime.

To cope with the rising design complexity a VP-based design flow is employed. In contrast to a traditional design flow which first builds the HW and then the SW, a VP-based design flow enables HW/SW Co-Design by leveraging the VP for early SW development and as reference model for the subsequent design flow steps. However, this modern VP-based design flow still has weaknesses, in particular due to the significant manual effort involved for verification and analysis tasks which is both time consuming and error prone. This book proposed several novel approaches to strongly enhance the VP-based design flow and with this provides contributions to every main step in a VP-based design flow. In particular, the book contributions have been divided into four areas which will be briefly summarized in the following:

1. VP Modeling
2. VP Verification
3. VP-based Approaches for SW Verification and Analysis
4. RTL/TLM Correspondence Analysis

The first contribution is an open-source RISC-V VP implemented in SystemC TLM. RISC-V is an open-source ISA that recently gained huge popularity in both academia and industry. The VP provides a 32/64 bit RISC-V core with an essential set of peripherals and support for multi-core simulations. Being implemented in SystemC TLM, the VP allows to leverage cutting-edge SystemC-based modeling techniques and provides the foundation for supporting system-level use cases such as design space exploration and timing/power evaluation in the RISC-V world. An

V. Herdt et al., *Enhanced Virtual Prototyping*,
https://doi.org/10.1007/978-3-030-54828-5_9

efficient core timing model was integrated into the VP to enable fast and accurate performance evaluation for RISC-V based systems. In summary, the VP allows a significantly faster simulation compared to RTL, while being more accurate than existing RISC-V simulators.

The second contribution improves the verification flow for VPs by considering novel formal verification methods and advanced automated coverage-guided testing techniques tailored for SystemC-based designs. In particular, new formal techniques based on symbolic simulation have been proposed that can efficiently deal with a large number of inputs as well as different process scheduling orders of SystemC. Albeit improving the state-of-the-art on formal verification for SystemC, formal approaches are still susceptible to state space explosion and hence alternative approaches are still important. Therefore, this book also considered advanced coverage-guided testing techniques that rely on testcase generation and simulation. Compared to the existing simulation-based verification flow this book investigates stronger coverage metrics based on *Data Flow Testing* (DFT) as well as advanced automated testcase generation and refinement techniques based on *Coverage-guided Fuzzing* (CGF).

The third contribution is coverage-guided approaches that improve the VP-based SW verification and analysis. The first proposed approach integrates concolic testing into the VP to generate a comprehensive SW testsuite by leveraging formal methods. The approach is very effective in maximizing the SW path coverage by relying on symbolic constraints but at the same time is also susceptible to scalability problems. Therefore, this book proposed a second SW verification approach that leverages state-of-the-art CGF in combination with VPs. To guide the fuzzing process and further improve the verification results the coverage from the embedded SW is combined with the coverage of the SystemC-based peripherals of the VP. Besides SW verification, this book proposed novel VP-based approaches tailored for validation of SW-based power management strategies. They leverage constrained random techniques and an automated coverage refinement loop for the generation of a large testsuite with different application workload characteristics.

The fourth and final contribution of this book are two approaches that perform a correspondence analysis between RTL and TLM. The first approach is an automated TLM-to-RTL property refinement to avoid the error prone and time consuming manual transformation of TLM properties into RTL properties. The second approach performs an RTL-to-TLM fault correspondence analysis to identify corresponding TLM errors for transient bit flips at RTL and thereby improving the accuracy of a VP-based error effect simulation.

All approaches have been implemented, discussed, and extensively evaluated with several experiments.[1] In summary, these contributions significantly enhance the VP-based design flow by drastically improving the verification quality and at the same time reduce the overall verification effort due to the extensive automatization.

[1]For our most recent VP-based approaches visit www.systemc-verification.org.

Outlook The obtained results clearly demonstrate the effectiveness and benefits of the proposed approaches. Nonetheless, there is still room for additional improvement and extension. In particular, the following four directions are very promising and important to further boost and complement this book's work:

- Increase the scalability of formal verification methods for SystemC further to mitigate state space explosion even stronger and enable application of formal techniques for the complete VP. The proposed *Compiled Symbolic Simulation* (CSS) technique provides a strong foundation to pursue this goal. Moreover, as an intermediate step, scalability can be improved by efficiently combining symbolic simulation with coverage-guided testing techniques to create a unified verification approach. This unified approach can seamlessly and intelligently switch between different formal and simulation-based verification techniques to provide a deep and broad cover of the state space more efficiently. Formal verification of SW has conceptually similar challenges and thus can benefit from these solutions as well.
- Investigate a fully automated and scalable cross-level methodology that enables to utilize VP verification results and provide them as verification-IP for RTL (i.e. the subsequent design flow steps). A verification-IP can, for example, be a pre-defined testsuite (that is designed to maximize specific coverage metrics) or a testsuite generator (e.g. leveraging constrained random or fuzzing techniques). A pre-defined testsuite is particularly suited for deep and positive testing, i.e. check that the expected behavior is available, while a testsuite generator is particularly effective for broad and negative testing, i.e. check that no additional behavior is accidentally implemented. Expected results can be encoded in the testbench or checked through a cross-level co-simulation. Alternatively, the verification-IP can also be a property set derived from the VP, which would enable formal reasoning but also be susceptible for scalability issues. The proposed automatic TLM-to-RTL property refinement approach is a step in the latter direction.
- Consider security aspects in a VP-based design flow to enable early and accurate evaluation of security policies that reason about data integrity and confidentiality aspects as well as side-channels. This is an important aspect, in particular for the upcoming highly connected IoT (*Internet-of-Things*) embedded devices that is complementary to the verification techniques proposed in this book. Moreover, comprehensive verification helps to mitigate security vulnerability risks by avoiding bugs such as buffer overflows that can be exploited by an attacker.
- Leverage and integrate verification techniques for SystemC AMS (*Analog/Mixed Signal*) to provide a unified verification environment that efficiently supports (inter-operating) analog and digital components (in combination with the SW). This is another complementary direction that broadens the verification scope to highly heterogeneous embedded systems with AMS components.

References

1. Accellera Systems Initiative. SystemC (2016). http://www.systemc.org
2. A. Adir, E. Almog, L. Fournier, E. Marcus, M. Rimon, M. Vinov, A. Ziv, Genesys-pro: innovations in test program generation for functional processor verification. Des. Test Comput. **21**(2), 84–93 (2004)
3. AFL-unicorn. Accessed 28 June 2019. https://github.com/Battelle/afl-unicorn
4. M. Aharoni, S. Asaf, L. Fournier, A. Koifman, R. Nagel, FPGEN—a test generation framework for datapath floating-point verification, in *Proceedings of the 8th IEEE International High-Level Design Validation and Test Workshop* (2003), pp. 17–22
5. A. Ahmed, F. Farahmandi, P. Mishra, Directed test generation using concolic testing on RTL models, in *Proceedings of the 2018 Design, Automation and Test in Europe Conference and Exhibition (DATE)* (2018), pp.1538–1543
6. S. Ahn, S. Malik, Automated firmware testing using firmware-hardware interaction patterns, in *Proceedings of the 2014 International Conference on Hardware/Software Codesign and System Synthesis* (2014), pp. 25:1–25:10
7. R.T. Alexander, J. Offutt, A. Stefik, Testing coupling relationships in object-oriented programs. STVR **20**(4), 291–327 (2010)
8. V. Alimi, S. Vernois, C. Rosenberger, Analysis of embedded applications by evolutionary fuzzing, in *Proceedings of the 2014 International Conference on High Performance Computing and Simulation (HPCS)* (2014), pp. 551–557
9. American fuzzy lop (2018). http://lcamtuf.coredump.cx/afl/
10. S. Anand, C.S. Păsăreanu, W. Visser, Symbolic execution with abstract subsumption checking, in *International SPIN Workshop on Model Checking of Software* (Springer, Berlin, 2006), pp. 163–181
11. B. Bailey, Power limits of EDA (2016). http://semiengineering.com/power-limits-of-eda
12. B. Bailey, G. Martin, A. Piziali, *ESL Design and Verification: A Prescription for Electronic System Level Methodology* (Morgan Kaufmann/Elsevier, New York, 2007)
13. F. Balarin, R. Passerone, Specification, synthesis, and simulation of transactor processes. IEEE Trans. Comput. Aided Des. Integr. Circuits Syst. **26**(10), 1749–1762 (2007)
14. G. Beltrame, C. Bolchini, A. Miele, Multi-level fault modeling for transaction-level specifications, in *Proceedings of the 19th ACM Great Lakes symposium on VLSI* (2009), pp. 87–92
15. J. Bennett, P. Dabbelt, C. Garlati, G.S. Madhusudan, T. Mudge, D. Patterson, Embench: an evolving benchmark suite for embedded iot computers from an academic-industrial cooperative (2019). https://content.riscv.org/wp-content/uploads/2019/06/9.25-Embench-RISC-V-Workshop-Patterson-v3.pdf. Accessed 07 juli 2019

V. Herdt et al., *Enhanced Virtual Prototyping*,
https://doi.org/10.1007/978-3-030-54828-5

16. A. Biere, D. Kroening, G. Weissenbacher, C. Wintersteiger, *Digitaltechnik—Eine Praxisnahe Einführung* (Springer, Berlin, 2008)
17. Big tech players start to adopt the risc-v chip architecture (2017). https://www.tomshardware.com/news/big-tech-players-risc-v-architecture,36011.html
18. N. Binkert, B. Beckmann, G. Black, S.K. Reinhardt, A. Saidi, A. Basu, J. Hestness, D.R. Hower, T. Krishna, S. Sardashti, R. Sen, K. Sewell, M. Shoaib, N. Vaish, M.D. Hill, D.A. Wood, The gem5 simulator. SIGARCH Comput. Archit. News **39**(2), 1–7 (2011)
19. D. Black, J. Donovan, B. Bunton, A. Keist, *SystemC: From the Ground Up*, 2nd edn. (Springer, New York, 2009)
20. N. Blanc, D. Kroening, Race analysis for SystemC using model checking. TODAES **15**(3), 21:1–21:32 (2010)
21. N. Bombieri, F. Fummi, G. Pravadelli, Incremental ABV for functional validation of TL-to-RTL design refinement, in *Proceedings of the 2007 Design, Automation and Test in Europe Conference and Exhibition* (2007), pp. 882–887
22. N. Bombieri, F. Fummi, G. Pravadelli, A mutation model for the SystemC TLM 2.0 communication interfaces, in *Proceedings of the 2008 Design, Automation and Test in Europe* (2008), pp. 396–401
23. N. Bombieri, F. Fummi, G. Pravadelli, On the mutation analysis of SystemC TLM-2.0 standard, in *MTV Workshop* (2009), pp. 32–37
24. I. Böhm, B. Franke, N. Topham, Cycle-accurate performance modelling in an ultra-fast just-in-time dynamic binary translation instruction set simulator, in *Proceedings of the 2010 International Conference on Embedded Computer Systems: Architectures, Modeling and Simulation* (2010), pp. 1–10
25. N. Bombieri, F. Fummi, V. Guarnieri, FAST: an RTL fault simulation framework based on RTL-to-TLM abstraction. J. Electron. Test. **28**(4), 495–510 (2012)
26. O. Bringmann, W. Ecker, A. Gerstlauer, A. Goyal, D. Mueller-Gritschneder, P. Sasidharan, S. Singh, The next generation of virtual prototyping: Ultra-fast yet accurate simulation of HW/SW systems, in *Proceedings of the 2015 Design, Automation and Test in Europe* (2015), pp. 1698–1707
27. C. Cadar, D. Dunbar, D. Engler, KLEE: unassisted and automatic generation of high-coverage tests for complex systems programs, in *OSDI* (2008), pp. 209–224
28. D. Campana, A. Cimatti, I. Narasamdya, M. Roveri, An analytic evaluation of SystemC encodings in Promela, in *International SPIN Workshop on Model Checking of Software* (2011), pp. 90–107
29. B. Campbell, I. Stark, Randomised testing of a microprocessor model using SMT-solver state generation, in *FMICS*, ed. by F. Lang, F. Flammini (2014), pp. 185–199
30. S.K. Cha, T. Avgerinos, A. Rebert, D. Brumley, Unleashing mayhem on binary code, in *IEEE S&P* (2012), pp. 380–394
31. A. Charif, G. Busnot, R. Mameesh, T. Sassolas, N. Ventroux, Fast virtual prototyping for embedded computing systems design and exploration, in *RAPIDO Workshop* (2019), pp. 3:1–3:8
32. M. Chen, P. Mishra, Assertion-based functional consistency checking between TLM and RTL models, in *Proceedings of the 2013 26th International Conference on VLSI Design and 2013 12th International Conference on Embedded Systems* (2013), pp. 320–325
33. M. Chiang, T. Yeh, G. Tseng, A QEMU and SystemC-based cycle-accurate ISS for performance estimation on SoC development. IEEE Trans. Comput. Aided Des. Integr. Circuits Syst. **30**(4), 593–606 (2011)
34. V. Chipounov, V. Kuznetsov, G. Candea, S2E: a platform for in-vivo multi-path analysis of software systems, in *ASPLOS* (2011), pp. 265–278
35. H. Cho, S. Mirkhani, C.Y. Cher, J.A. Abraham, S. Mitra, Quantitative evaluation of soft error injection techniques for robust system design, in *Proceedings of the 55th Annual Design Automation Conference* (2013), pp. 1–10
36. C.-N. Chou, Y.-S. Ho, C. Hsieh, C.-Y. Huang, Symbolic model checking on SystemC designs, in *DAC Design Automation Conference 2012* (2012), pp. 327–333

37. C. Chou, C. Chu, C. Huang, Conquering the scheduling alternative explosion problem of SystemC symbolic simulation, in *Proceedings of the 2013 IEEE/ACM International Conference on Computer-Aided Design (ICCAD)* (2013), pp. 685–690
38. A. Cimatti, I. Narasamdya, M. Roveri, Software model checking SystemC. IEEE Trans. Comput. Aided Des. Integr. Circuits Syst. **32**(5), 774–787 (2013)
39. N. Corteggiani, G. Camurati, A. Francillon, Inception: system-wide security testing of real-world embedded systems software, in *USENIX Security* (2018), pp. 309–326
40. P. Dasgupta, M.K. Srivas, R. Mukherjee, Formal hardware/software co-verification of embedded power controllers. IEEE Trans. Comput. Aided Des. Integr. Circuits Syst. **33**(12), 2025–2029 (2014)
41. D. Davidson, B. Moench, T. Ristenpart, S. Jha, FIE on firmware: finding vulnerabilities in embedded systems using symbolic execution, in *USENIX Security* (2013), pp. 463–478
42. DBT-RISE. https://github.com/Minres/DBT-RISE-Core. Accessed 04 March 2019
43. R.A. DeMillo, R.J. Lipton, F.G. Sayward, Hints on test data selection: help for the practicing programmer. Computer **11**(4), 34–41 (1978)
44. L. De Moura, N. Bjørner, Z3: an efficient SMT solver, in *International conference on Tools and Algorithms for the Construction and Analysis of Systems* (2008), pp. 337–340. https://github.com/Z3Prover/z3
45. G. Denaro, A. Margara, M. Pezzè, M. Vivanti, Dynamic data flow testing of object oriented systems, in *Proceedings of the 2015 IEEE/ACM 37th IEEE International Conference on Software Engineering* (2015), pp. 947–958
46. T. De Schutter, *Better Software. Faster!: Best Practices in Virtual Prototyping* (Synopsys, Mountain View, 2014)
47. R. Drechsler, D. Große, Reachability analysis for formal verification of SystemC, in *Proceedings Euromicro Symposium on Digital System Design. Architectures, Methods and Tools* (2002), pp. 337–340
48. W. Ecker, V. Esen, M. Hull, Implementation of a transaction level assertion framework in systemc, in *Proceedings of the 2007 Design, Automation and Test in Europe* (2007), pp. 1–6
49. W. Ecker, V. Esen, T. Steininger, M. Velten, Requirements and concepts for transaction level assertion refinement, in *Embedded System Design: Topics, Techniques and Trends* (2007), pp. 1–14
50. Embench: a modern embedded benchmark suite. https://www.embench.org/
51. Embench: open benchmarks for embedded platforms. Accessed 07 Juli 2019. https://github.com/embench/embench-iot
52. R. Emek, I. Jaeger, Y. Naveh, G. Bergman, G. Aloni, Y. Katz, M. Farkash, I. Dozoretz, A. Goldin, X-gen: a random test-case generator for systems and SOCS, in *Proceedings of the 7th IEEE International High-Level Design Validation and Test Workshop, 2002* (2002), pp. 145–150
53. S. Evangelista, C. Pajault, Solving the ignoring problem for partial order reduction. STTT **12**(2), 155–170 (2010)
54. L. Ferro, L. Pierre, ISIS: runtime verification of tlm platforms, in *FDL* (2009), pp. 1–6
55. L. Ferro, L. Pierre, Formal semantics for PSL modeling layer and application to the verification of transactional models, in *Proceedings of the 2010 Design, Automation and Test in Europe* (2010), pp. 1207–1212
56. S. Fine, A. Ziv, Coverage directed test generation for functional verification using bayesian networks, in *Proceedings of the 40th annual Design Automation Conference* (2003), pp. 286–291
57. C. Flanagan, P. Godefroid, Dynamic Partial-Order Reduction for model checking software, in *POPL* (2005), pp. 110–121
58. Forvis: a formal RISC-V ISA specification. https://github.com/rsnikhil/RISCV-ISA-Spec
59. H. Foster, Applied assertion-based verification: an industry perspective. Found. Trends Electron. Des. Autom. **3**(1), 1–95 (2009)
60. FreeRTOS. https://www.freertos.org/

61. FreeRTOS TCP/IP stack vulnerabilities put a wide range of devices at risk of compromise: from smart homes to critical infrastructure systems. https://blog.zimperium.com/freertos-tcpip-stack-vulnerabilities-put-wide-range-devices-risk-compromise-smart-homes-critical-infrastructure-systems
62. P. Godefroid, *Partial-Order Methods for the Verification of Concurrent Systems: An Approach to the State-Explosion Problem* (Springer, New York, 1996)
63. P. Godefroid, N. Klarlund, K. Sen, Dart: directed automated random testing, in *Proceedings of the 2005 ACM SIGPLAN conference on Programming language Design and Implementation* (2005), pp. 213–223
64. P. Godefroid, M.Y. Levin, D. A. Molnar, Automated whitebox fuzz testing, in *NDSS* (2008)
65. P. Godefroid, H. Peleg, R. Singh, Learn and fuzz: machine learning for input fuzzing, in *ASE* (2017), pp. 50–59
66. GreenReg. http://www.greensocs.com/projects/GreenReg
67. GRIFT—galois RISC-V ISA formal tools. https://github.com/GaloisInc/grift
68. GRLIB IP library. http://www.gaisler.com/index.php/products/ipcores/soclibrary
69. A. Griesmayer, S. Staber, R. Bloem, Automated fault localization for C programs. Electr. Notes Theor. Comput. Sci. **174**(4), 95–111 (2007)
70. D. Große, R. Drechsler, CheckSyC: an efficient property checker for RTL SystemC designs, in *ISCAS* (2005), pp. 4167–4170
71. D. Große, R. Drechsler, *Quality-Driven SystemC Design* (Springer, New York, 2010)
72. D. Große, H.M. Le, R. Drechsler, Proving transaction and system-level properties of untimed SystemC TLM designs, in *MEMOCODE* (2010), pp. 113–122
73. D. Große, H.M. Le, M. Hassan, R. Drechsler, Guided lightweight software test qualification for IP integration using virtual prototypes, in *ICCD* (2016), pp. 606–613
74. T. Grotker, *System Design with SystemC* (Kluwer Academic, Berlin, 2002)
75. K. Grüttner, P.A. Hartmann, K. Hylla, S. Rosinger, W. Nebel, F. Herrera, E. Villar, C. Brandolese, W. Fornaciari, G. Palermo, C. Ykman-Couvreur, D. Quaglia, F. Ferrero, R. Valencia, The COMPLEX reference framework for HW/SW co-design and power management supporting platform-based design-space exploration. MICPRO **37**(8, Part C), 966–980 (2013)
76. K. Grüttner, R. Görgen, S. Schreiner, F. Herrera, P. Peñil, J. Medina, E. Villar, G. Palermo, W. Fornaciari, C. Brandolese, D. Gadioli, E. Vitali, D. Zoni, S. Bocchio, L. Ceva, P. Azzoni, M. Poncino, S. Vinco, E. Macii, S. Cusenza, J. Favaro, R. Valencia, I. Sander, K. Rosvall, N. Khalilzad, D. Quaglia, CONTREX: design of embedded mixed-criticality CONTRol systems under consideration of EXtra-functional properties. MICPRO **51**, 39–55 (2017)
77. F. Haedicke, H.M. Le, D. Große, R. Drechsler, CRAVE: an advanced constrained random verification environment for SystemC, in *SoC* (2012), pp. 1–7
78. R.G. Hamlet, Testing programs with the aid of a compiler. IEEE Trans. Soft. Eng. **3**(4), 279–290 (1977)
79. M. Hassan, V. Herdt, H.M. Le, M. Chen, D. Große, R. Drechsler, Data flow testing for virtual prototypes, in *Proceedings of the 2017 Design, Automation and Test in Europe* (2017), pp. 380–385
80. M. Hassan, V. Herdt, H.M. Le, D. Große, R. Drechsler, Early SoC security validation by VP-based static information flow analysis, in *Proceedings of the 2017 IEEE/ACM International Conference on Computer-Aided Design (ICCAD)* (2017), pp. 400–407
81. P. Herber, The rescue approach—towards compositional hardware/software co-verification, in *HPCC,CSS,ICESS* (2014), pp. 721–724
82. P. Herber, J. Fellmuth, S. Glesner, Model checking SystemC designs using timed automata, in *CODES+ISSS* (2008), pp. 131–136
83. P. Herber, M. Pockrandt, S. Glesner, Transforming SystemC transaction level models into UPPAAL timed automata, in *MEMOCODE* (2011), pp. 161–170
84. P. Herber, M. Pockrandt, S. Glesner, State—a systemc to timed automata transformation engine, in *HPCC-CSS-ICESS* (2015), pp. 1074–1077
85. V. Herdt, *Complete Symbolic Simulation of SystemC Models: Efficient Formal Verification of Finite Non-Terminating Programs*, 1st edn. (Springer, Berlin, 2016)

86. V. Herdt, H.M. Le, R. Drechsler, Verifying SystemC using stateful symbolic simulation, in *Proceedings of the 2015 52nd ACM/EDAC/IEEE Design Automation Conference (DAC)* (2015), pp. 49:1–49:6
87. V. Herdt, H.M. Le, D. Große, R. Drechsler, Boosting sequentialization-based verification of multi-threaded C programs via symbolic pruning of redundant schedules, in *ATVA* (2015), pp. 228–233
88. V. Herdt, H.M. Le, D. Große, R. Drechsler, Compiled symbolic simulation for SystemC, in *Proceedings of the 2016 IEEE/ACM International Conference on Computer-Aided Design (ICCAD)* (2016), pp. 52:1–52:8
89. V. Herdt, H.M. Le, D. Große, R. Drechsler, On the application of formal fault localization to automated RTL-to-TLM fault correspondence analysis for fast and accurate VP-based error effect simulation—a case study, in *FDL* (2016), pp. 1–8
90. V. Herdt, H.M. Le, D. Große, R. Drechsler, ParCoSS: efficient parallelized compiled symbolic simulation, in *CAV* (2016), pp. 177–183
91. V. Herdt, H.M. Le, D. Große, R. Drechsler, Towards early validation of firmware-based power management using virtual prototypes: a constrained random approach, in *FDL* (2017), pp. 1–8
92. V. Herdt, H.M. Le, D. Große, R. Drechsler, *Intermediate Verification Language for SystemC—Language Reference Manual* (University of Bremen, Bremen, 2018)
93. V. Herdt, H.M. Le, D. Große, R. Drechsler, On the application of formal fault localization to automated RTL-to-TLM fault correspondence analysis for fast and accurate VP-based error effect simulation—a case study, in *Selected Contributions from FDL 2016*, ed. by F. Fummi, R. Wille (Springer, Berlin, 2018), pp. 39–58.
94. V. Herdt, H.M. Le, D. Große, R. Drechsler, Towards fully automated TLM-to-RTL property refinement, in *Proceedings of the 2018 Design, Automation and Test in Europe* (2018), pp. 1508–1511
95. V. Herdt, D. Große, H.M. Le, R. Drechsler, Extensible and configurable RISC-V based virtual prototype, in *Selected Contributions from FDL* (Springer, Berlin, 2018)
96. V. Herdt, D. Große, H.M. Le, R. Drechsler, Extensible and configurable RISC-V based virtual prototype, in *FDL* (2018), pp. 5–16
97. V. Herdt, H.M. Le, D. Große, R. Drechsler, Combining sequentialization-based verification of multi-threaded C programs with symbolic partial order reduction. STTT **21**(5), 545–565 (2019)
98. V. Herdt, H.M. Le, D. Große, R. Drechsler, Maximizing power state cross coverage in firmware-based power management, in *ASP-DAC* (2019), pp. 335–340
99. V. Herdt, H.M. Le, D. Große, R. Drechsler, Towards early validation of firmware-based power management using virtual prototypes: a constrained random approach, in *Selected Contributions from FDL 2017*, ed. by D. Große, S. Vinco, H. Patel (Springer, New York, 2019), pp. 25–44
100. V. Herdt, H.M. Le, D. Große, R. Drechsler, Verifying SystemC using intermediate verification language and stateful symbolic simulation. IEEE Trans. Comput. Aided Des. Integr. Circuits Syst. **38**(7), 1359–1372 (2019)
101. V. Herdt, D. Große, H.M. Le, R. Drechsler, Early concolic testing of embedded binaries with virtual prototypes: a RISC-V case study, in *Proceedings of the 2019 56th ACM/IEEE Design Automation Conference (DAC)* (2019), pp. 188:1–188:6
102. V. Herdt, D. Große, H.M. Le, R. Drechsler, Verifying instruction set simulators using coverage-guided fuzzing, in *Proceedings of the 2019 Design, Automation and Test in Europe* (2019), pp. 360–365
103. V. Herdt, D. Große, R. Drechsler, Closing the RISC-V compliance gap: looking from the negative testing side, in *Design Automation Conference* (2020)
104. V. Herdt, D. Große, R. Drechsler, Fast and accurate performance evaluation for RISC-V using virtual prototypes, in *Proceedings of the 2020 Design, Automation and Test in Europe* (2020)
105. V. Herdt, D. Große, R. Drechsler, Towards specification and testing of RISC-V ISA compliance, in *Proceedings of the 2020 Design, Automation and Test in Europe* (2020)

106. V. Herdt, D. Große, P. Pieper, R. Drechsler, RISC-V based virtual prototype: an extensible and configurable platform for the system-level. J. Syst. Archit. Embedded Soft. Des., **109**, 101756 (2020). https://www.sciencedirect.com/science/article/abs/pii/S1383762120300503
107. V. Herdt, D. Große, J. Wloka, T. Güneysu, R. Drechsler, Verification of embedded binaries using coverage-guided fuzzing with SystemC-based virtual prototypes, in *ACM Great Lakes Symposium on VLSI* (2020)
108. HiFive1. https://www.sifive.com/boards/hifive1
109. G.J. Holzmann, The model checker SPIN. IEEE Trans. Soft. Eng. **23**(5), 279–295 (1997)
110. A. Horn, M. Tautschnig, C.G. Val, L. Liang, T. Melham, J. Grundy, D. Kroening, Formal co-validation of low-level hardware/software interfaces, in *FMCAD* (2013), pp. 121–128
111. B. Huang, S. Ray, A. Gupta, J.M. Fung, S. Malik, Formal security verification of concurrent firmware in SoCs using instruction-level abstraction for hardware, in *Proceedings of the 55th Annual Design Automation Conference* (2018), pp. 91:1–91:6
112. HW-SW SystemC co-simulation SoC validation platform. Technical report, TU Braunschweig (2012)
113. IEEE Standard 1666, *IEEE Standard SystemC Language Reference Manual* (2011)
114. Intel Corporation, *IA-32 Architecture Software Developer's Manual* (2003)
115. C. Ioannides, K.I. Eder, Coverage-directed test generation automated by machine learning—a review. TODAES **17**(1), 7:1–7:21 (2012)
116. C. Ioannides, G. Barrett, K. Eder, Feedback-based coverage directed test generation: an industrial evaluation, in *HVC*, ed. by S. Barner, I. Harris, D. Kroening, O. Raz (2011), pp. 112–128
117. N. Kamel, J.-L. Lanet, Analysis of http protocol implementation in smart card embedded web server, in *IJINS* (2013), pp. 417–428
118. M. Kammerstetter, C. Platzer, W. Kastner, Prospect: peripheral proxying supported embedded code testing, in *ASIA CCS* (2014), pp. 329–340
119. D. Karlsson, P. Eles, Z. Peng, Formal verification of systemc designs using a petri-net based representation, in *Proceedings of the 2006 Design, Automation and Test in Europe* (2006), pp. 1228–1233
120. J. Karmann, W. Ecker, The semantic of the power intent format UPF: consistent power modeling from system level to implementation, in *PATMOS* (2013), pp. 45–50
121. Y. Katz, M. Rimon, A. Ziv, Generating instruction streams using abstract CSP, in *Proceedings of the 2012 Design, Automation and Test in Europe* (2012), pp. 15–20
122. J.C. King, Symbolic execution and program testing. Commun. Assoc. Comput. Mach. **19**(7), 385–394 (1976)
123. T. Kogel, Peripheral modeling for platform driven ESL design, in *Platform Based Design at the Electronic System Level* (Springer, Netherlands, 2006), pp. 71–85
124. J. Kong, B. Yoo, D. Song, H.J. Nam, J. Hwang, J. Kim, S. Lee, S. Eo, S. Yoo, K. Choi, H. Jin, J. Kim, S. Lee, S. Hong, Creation and utilization of a virtual platform for embedded software optimization:: an industrial case study, in *CODES+ISSS* (2006), pp. 235–240
125. K. Koscher, A. Czeskis, F. Roesner, S. Patel, T. Kohno, S. Checkoway, D. McCoy, B. Kantor, D. Anderson, H. Shacham, S. Savage, Experimental security analysis of a modern automobile, in *IEEE S&P* (2010), pp. 447–462
126. A. Kramer, M. Vaupel, Virtual platforms for automotive: Use cases, benefits and challenges (2014). https://dvcon-europe.org/sites/dvcon-europe.org/files/archive/2014/proceedings/T01_tutorial_part3.pdf
127. D. Kroening, N. Sharygina, Formal verification of SystemC by automatic hardware/software partitioning, in *MEMOCODE* (2005), pp. 101–110
128. S. Kundu, M. Ganai, R. Gupta, Partial order reduction for scalable testing of SystemC TLM designs, in *DAC* (2008), pp. 936–941
129. V. Kuznetsov, J. Kinder, S. Bucur, G. Candea, Efficient state merging in symbolic execution, in *Proceedings of the 2005 ACM SIGPLAN conference on Programming language Design and Implementation* (2012), pp. 193–204

130. J.W. Laski, B. Korel, A data flow oriented program testing strategy. IEEE Trans. Soft. Eng. **9**(3), 347–354 (1983)
131. J. Laurent, N. Julien, E. Senn, E. Martin, Functional level power analysis: an efficient approach for modeling the power consumption of complex processors, in *Proceedings of the 2004 Design, Automation and Test in Europe* (2004)
132. H.M. Le, R. Drechsler, CRAVE 2.0: the next generation constrained random stimuli generator for SystemC, in *DVCon* (2014)
133. H.M. Le, D. Große, R. Drechsler, Automatic TLM fault localization for SystemC. IEEE Trans. Comput. Aided Des. Integr. Circuits Syst. **31**(8), 1249–1262 (2012)
134. H.M. Le, D. Große, V. Herdt, R. Drechsler, Verifying SystemC using an intermediate verification language and symbolic simulation, in *DAC* (2013), pp. 116:1–116:6
135. H.M. Le, V. Herdt, D. Große, R. Drechsler, Towards formal verification of real-world SystemC TLM peripheral models—a case study, in *Proceedings of the 2016 Design, Automation and Test in Europe* (2016), pp. 1160–1163
136. H.M. Le, V. Herdt, D. Große, R. Drechsler, Resiliency evaluation via symbolic fault injection on intermediate code, in *Proceedings of the 2018 Design, Automation and Test in Europe* (2018), pp. 845–850
137. D. Lee, J. Na, A novel simulation fault injection method for dependability analysis. Des. Test Comput. **26**(6), 50–61 (2009)
138. H. Lee, K. Choi, K. Chung, J. Kim, K. Yim, Fuzzing can packets into automobiles, in *AINA* (2015), pp. 817–821
139. R. Leupers, F. Schirrmeister, G. Martin, T. Kogel, R. Plyaskin, A. Herkersdorf, M. Vaupel, Virtual platforms: Breaking new grounds, in *Proceedings of the 2012 Design, Automation and Test in Europe* (2012), pp. 685–690
140. M.L. Li, P. Ramachandran, U.R. Karpuzcu, S.K.S. Hari, S.V. Adve, Accurate microarchitecture-level fault modeling for studying hardware faults, in *HPCA* (2009), pp. 105–116
141. Libfuzzer—a library for coverage-guided fuzz testing (2018). https://llvm.org/docs/LibFuzzer.html
142. B. Lin, K. Cong, Z. Yang, Z. Liao, T. Zhan, C. Havlicek, F. Xie, Concolic testing of SystemC designs, in *ISQED* (2018), pp. 1–7
143. P. Lisherness, K.-T. (Tim) Cheng, SCEMIT: a SystemC error and mutation injection tool, in *DAC* (2010), pp. 228–233
144. L. Liu, D. Sheridan, W. Tuohy, and S. Vasudevan, A technique for test coverage closure using goldmine. IEEE Trans. Comput. Aided Des. Integr. Circuits Syst. **31**(5), 790–803 (2012)
145. K. Lu, D. Müller-Gritschneder, U. Schlichtmann, Accurately timed transaction level models for virtual prototyping at high abstraction level, in *Proceedings of the 2012 Design, Automation and Test in Europe* (2012), pp. 135–140
146. T. Ludwig, J. Urdahl, D. Stoffel, W. Kunz, Properties first—correct-by-construction RTL design in system-level design flows. IEEE Trans. Comput. Aided Des. Integr. Circuits Syst. **39**, 3093–3106 (2019). https://ieeexplore.ieee.org/document/8759950
147. W. Ma, A. Forin, J. Liu, Rapid prototyping and compact testing of CPU emulators, in *RSP* (2010), pp. 1–7
148. K. Marquet, M. Moy, Pinavm: a systemc front-end based on an executable intermediate representation, in *EMSOFT*, pp. 79–88 (2010)
149. L. Martignoni, R. Paleari, G.F. Roglia, D. Bruschi, Testing CPU emulators, in *ISSTA* (2009), pp. 261–272
150. O. Mbarek, A. Pegatoquet, M. Auguin, Using unified power format standard concepts for power-aware design and verification of systems-onchip at transaction level. IET Circuits Devices Syst. **6**(5), 287–296 (2012)
151. Members at a glance (2019). https://riscv.org/members-at-a-glance/
152. Microsoft security development lifecycle (2018). https://www.microsoft.com/en-us/sdl/process/verification.aspx

153. A. Miele, A fault-injection methodology for the system-level dependability analysis of multiprocessor embedded systems. MICPRO **38**(6), 567–580 (2014)
154. B.P. Miller, L. Fredriksen, B. So, An empirical study of the reliability of unix utilities. Commun. ACM **33**(12), 32–44 (1990)
155. M. Moy, F. Maraninchi, L. Maillet-Contoz, Lussy: an open tool for the analysis of systems-on-a-chip at the transaction level. ACSD **10**(2–3), 73–104 (2005)
156. M. Muench, J. Stijohann, F. Kargl, A. Francillon, D. Balzarotti, What you corrupt is not what you crash: Challenges in fuzzing embedded devices, in *NDSS* (2018)
157. R. Mukherjee, M. Purandare, R. Polig, D. Kroening, Formal techniques for effective co-verification of hardware/software co-designs, in *DAC* (2017), pp. 35:1–35:6
158. C. Mulliner, N. Golde, J.-P. Seifert, Sms of death: from analyzing to attacking mobile phones on a large scale, in *USENIX Security* (2011), pp. 24–24
159. Neural fuzzing: applying DNN to software security testing. https://www.microsoft.com/en-us/research/blog/neural-fuzzing/
160. J.H. Oetjens, N. Bannow, M. Becker, O. Bringmann, A. Burger, M. Chaari, S. Chakraborty, R. Drechsler, W. Ecker, K. Grüttner, T. Kruse, C. Kuznik, H.M. Le, A. Mauderer, W. Müller, D. Müller-Gritschneder, F. Poppen, H. Post, S. Reiter, W. Rosenstiel, S. Roth, U. Schlichtmann, A. von Schwerin, B.A. Tabacaru, A. Viehl, Safety evaluation of automotive electronics using virtual prototypes: state of the art and research challenges, in *DAC* (2014), pp. 1–6
161. M.F.S. Oliveira et al., The system verification methodology for advanced TLM verification, in *CODES+ISSS* (2012), pp. 313–322
162. G. Onnebrink, R. Leupers, G. Ascheid, S. Schürmans, Black box ESL power estimation for loosely-timed TLM models, in *Proceedings of the 2016 International Conference on Embedded Computer Systems: Architectures, Modeling and Simulation* (2016), pp. 366–371
163. OSCI, *OSCI TLM-2.0 Language Reference Manual* (2009)
164. Oss-fuzz—continuous fuzzing for open source software (2018). https://github.com/google/oss-fuzz
165. S. Ottlik, S. Stattelmann, A. Viehl, W. Rosenstiel, O. Bringmann, Context-sensitive timing simulation of binary embedded software, in *CASES* (2014), pp. 1–10
166. K. Pattabiraman, N.M. Nakka, Z.T. Kalbarczyk, R.K. Iyer, Symplfied: symbolic program-level fault injection and error detection framework. IEEE Trans. Comput. **62**(11), 2292–2307 (2013)
167. Pcgamesn article: Nvidia's 36-module research chip is paving the way to multi-gpu graphics cards (2019). https://riscv.org/2019/04/pcgamesn-article-nvidias-36-module-research-chip-is-paving-the-way-to-multi-gpu-graphics-cards/
168. J. Perez, M. Azkarate-askasua, A. Perez, Codesign and simulated fault injection of safety-critical embedded systems using SystemC, in *EDCC* (2010), pp. 221–229
169. P. Pieper, V. Herdt, D. Große, R. Drechsler, Dynamic information flow tracking for embedded binaries using SystemC-based virtual prototypes, in *Design Automation Conference* (2020)
170. L. Pierre, Z.B.H. Amor, Automatic refinement of requirements for verification throughout the SoC design flow, in *CODES+ISSS* (2013), pp. 29:1–29:10
171. S. Pinto, M.S. Hsiao, RTL functional test generation using factored concolic execution, in *ITC* (2017), pp. 1–10
172. M. Pockrandt, P. Herber, S. Glesner, Model checking a SystemC/TLM design of the AMBA AHB protocol, in *ESTIMedia* (2011), pp. 66–75
173. Porting FreeRTOS+TCP to a different microcontroller. https://www.freertos.org/FreeRTOS-Plus/FreeRTOS_Plus_TCP/Embedded_Ethernet_Porting.html
174. Project triforce: run afl on everything! Accessed 28 june 2019. https://www.nccgroup.trust/us/about-us/newsroom-and-events/blog/2016/june/project-triforce-run-afl-on-everything
175. S. Rapps, E.J. Weyuker, Selecting software test data using data flow information. IEEE Trans. Soft. Eng. **11**(4), 367–375 (1985)
176. Renode. https://renode.io/. Accessed 04 March 2019

177. S.K. Rethinagiri, O. Palomar, R. Ben Atitallah, S. Niar, O. Unsal, A.C. Kestelman, System-level power estimation tool for embedded processor based platforms, in *RAPIDO Workshop* (2014), pp. 5:1–5:8
178. RISC-V in NVIDIA (2017). https://riscv.org/wp-content/uploads/2017/05/Tue1345pm-NVIDIA-Sijstermans.pdf
179. RISC-V calling convention. https://riscv.org/wp-content/uploads/2015/01/riscv-calling.pdf. Accessed 13 May 2018
180. RISC-V Foundation. https://riscv.org/
181. RISC-V ISA tests. https://github.com/riscv/riscv-tests
182. RISC-V torture test generator. https://github.com/ucb-bar/riscv-torture
183. RISC-V virtual prototype. https://github.com/agra-uni-bremen/riscv-vp
184. RISCV sail model. https://github.com/rems-project/sail-riscv
185. RISCV-QEMU. https://github.com/riscv/riscv-qemu. Accessed 13 May 2018
186. RV8. https://rv8.io. Accessed 13 May 2018
187. P. Sayyah, M.T. Lazarescu, S. Bocchio, E. Ebeid, G. Palermo, D. Quaglia, A. Rosti, L. Lavagno, Virtual platform-based design space exploration of power-efficient distributed embedded applications. ACM TECS **14**(3), 49:1–49:25 (2015)
188. S. Schürmans, D. Zhang, D. Auras, R. Leupers, G. Ascheid, X. Chen, L. Wang, Creation of ESL power models for communication architectures using automatic calibration, in *DAC* (2013), pp. 1–6
189. T. Schuster, R. Meyer, R. Buchty, L. Fossati, M. Berekovic, Socrocket—A virtual platform for the European Space Agency's SoC development, in *ReCoSoC* (2014), pp. 1–7. http://github.com/socrocket
190. A. Sen, Concurrency-oriented verification and coverage of system-level designs. TODAES **16**(4), 37 (2011)
191. A. Sen, M.S. Abadir, Coverage metrics for verification of concurrent systemc designs using mutation testing, in *Proceedings of the Seventh IEEE International High-Level Design Validation and Test Workshop, 2002* (2010), pp. 75–81
192. K. Sen, D. Marinov, G. Agha, Cute: a concolic unit testing engine for C, in *SEN* (2005), pp. 263–272
193. R.A. Shafik, P. Rosinger, B.M. Al-Hashimi, SystemC-based minimum intrusive fault injection technique with improved fault representation, in *Proceedings of the 2008 14th IEEE International On-Line Testing Symposium* (2008), pp. 99–104
194. Y. Shoshitaishvili, R. Wang, C. Salls, N. Stephens, M. Polino, A. Dutcher, J. Grosen, S. Feng, C. Hauser, C. Krügel, G. Vigna, SOK: (state of) the art of war: offensive techniques in binary analysis, in *IEEE S&P* (2016), pp. 138–157
195. SiFive FE310-G000 Manual v2p3 (2017). Accessed 28 June 2019. https://sifive.cdn.prismic.io/sifive%2F4d063bf8-3ae6-4db6-9843-ee9076ebadf7_fe310-g000.pdf. Accessed 03 April 2019
196. V.V. Singh, A. Kumar, Cross coverage of power states, in *DVCon* (2016)
197. Spike RISC-V ISA simulator. https://github.com/riscv/riscv-isa-sim
198. S. Stattelmann, G. Gebhard, C. Cullmann, O. Bringmann, W. Rosenstiel, Hybrid source-level simulation of data caches using abstract cache models, in *Proceedings of the 2012 Design, Automation and Test in Europe* (2012), pp. 376–381
199. T. Steininger, Automated assertion transformation across multiple abstraction levels, PhD thesis (TU Munich, Germany, 2009)
200. M. Streubühr, R. Rosales, R. Hasholzner, C. Haubelt, J. Teich, ESL power and performance estimation for heterogeneous mpsocs using SystemC, in *FDL* (2011), pp. 1–8
201. T. Su, Z. Fu, G. Pu, J. He, Z. Su, Combining symbolic execution and model checking for data flow testing, in *Proceedings of the 2015 IEEE/ACM 37th IEEE International Conference on Software Engineering* (2015), pp. 654–665
202. S. Swan, J. Cornet, Beyond TLM 2.0: new virtual platform standards proposals from ST and Cadence, in *Presented at NASCUG at DAC* (2012), p. 18

203. Synopsys Inc. SystemC Modeling Library (SCML). https://www.synopsys.com/cgi-bin/slcw/kits/reg.cgi
204. D. Tabakov, M.Y. Vardi, Monitoring temporal SystemC properties, in *MEMOCODE* (2010), pp. 123–132
205. D. Tabakov, M. Vardi, G. Kamhi, E. Singerman, A temporal language for SystemC, in *FMCAD* (2008), pp. 1–9
206. D. Thach, Y. Tamiya, S. Kuwamura, A. Ike, Fast cycle estimation methodology for instruction-level emulator, in *Proceedings of the 2012 Design, Automation and Test in Europe* (2012), pp. 248–251
207. C. Traulsen, J. Cornet, M. Moy, F. Maraninchi, A SystemC/TLM semantics in Promela and its possible applications, in *International SPIN Workshop on Model Checking of Software* (2007), pp. 204–222
208. Triforceafl. Accessed 28 June 2019. https://github.com/nccgroup/TriforceAFL
209. J. Urdahl, D. Stoffel, W. Kunz, Path predicate abstraction for sound system-level models of rt-level circuit designs. IEEE Trans. Comput. Aided Des. Integr. Circuits Syst. **33**(2), 291–304 (2014)
210. A. Valmari, Stubborn sets for reduced state space generation, in *International Conference on Application and Theory of Petri Nets* (1989), pp. 1–22
211. F. van den Broek, B. Hond, A. Cedillo Torres, Security testing of gsm implementations, in *International Symposium on Engineering Secure Software and Systems* (2014), pp. 179–195
212. M.Y. Vardi, Formal techniques for SystemC verification, in *Proceedings of the 44th annual Design Automation Conference* (2007), pp. 188–192
213. M. Vivanti, A. Mis, A. Gorla, G. Fraser, Search-based data-flow test generation, in *Proceedings of the 2013 IEEE 24th International Symposium on Software Reliability Engineering (ISSRE)* (2013), pp. 370–379
214. H. Wagstaff, T. Spink, B. Franke, Automated ISA branch coverage analysis and test case generation for retargetable instruction set simulators, in *Proceedings of the 2014 International Conference on Compilers, Architecture and Synthesis for Embedded Systems* (2014), pp. 1–10
215. Z. Wang, J. Henkel, Fast and accurate cache modeling in source-level simulation of embedded software, in *Proceedings of the 2013 Design, Automation and Test in Europe* (2013), pp. 587–592
216. Z. Wang, K. Lu, A. Herkersdorf, An approach to improve accuracy of source-level tlms of embedded software, in *Proceedings of the 2011 Design, Automation and Test in Europe* (2011), pp. 1–6
217. B. Wang, Y. Xu, R. Hasholzner, C. Drewes, R. Rosales, S. Graf, J. Falk, M. Glaß, J. Teich, Exploration of power domain partitioning for application-specific SoCs in system-level design, in *MBMV Workshop* (2016), pp. 102–113
218. A. Waterman, K. Asanović, in *The RISC-V Instruction Set Manual*. Unprivileged ISA, vol. I (SiFive Inc. and CS Division/EECS Department, University of California, Berkeley, 2019)
219. A. Waterman, K. Asanović, *The RISC-V Instruction Set Manual*. Privileged Architecture (SiFive Inc. and CS Division/EECS Department, University of California, Berkeley, 2019)
220. Western digitals risc-v "SweRV" core design released for free (2019). https://www.anandtech.com/show/13964/western-digitals-riscv-swerv-core-released-for-free
221. W. Ye, N. Vijaykrishnan, M. Kandemir, M.J. Irwin, The design and use of simplepower: a cycle-accurate energy estimation tool, in *Proceedings of the 37th Annual Design Automation Conference* (2000), pp. 340–345
222. J. Yuan, C. Pixley, A. Aziz *Constraint-based Verification* (Springer, Berlin, 2006)
223. J. Zaddach, L. Bruno, A. Francillon, D. Balzarotti, AVATAR: a framework to support dynamic security analysis of embedded systems' firmwares, in *NDSS* (2014)
224. Zephyr Project. https://www.zephyrproject.org/

Index

V. Herdt et al., *Enhanced Virtual Prototyping*,
https://doi.org/10.1007/978-3-030-54828-5

Printed in the United States
by Baker & Taylor Publisher Services